We live in an old chaos of the sun.

Wallace Stevens

PENGUIN BOOKS

NOVACENE

'Leavened with wit and optimism . . . *Novacene* is the collected wisdom of an elder of our tribe which more than repays the short time it takes to read' Stephen Cave, *Financial Times*

'Astonishing . . . it's invigorating to spend time with such an unshackled brain' Frank Cottrell Boyce, *The Tablet*, Books of the Year

'This restlessly thoughtful and forward-looking book . . . is partly a defence of a lifetime's ideas, but mostly an argument about how AI is soon to overtake us – and what that means for our species' James McConnachie, *The Times*

'In a brief but thought-provoking book, the scientist who developed the "Gaia Theory" predicts that cyborgs may eventually evolve to supplant carbon-based humankind. But don't despair: the cyborgs, he suggests, might decide to keep people around as pets' *Economist*, Books of the Year

ABOUT THE AUTHOR

James Lovelock, who was elected a Fellow of the Royal Society in 1974, is the author of more than 200 scientific papers and the originator of the Gaia Hypothesis (now Gaia Theory). His many books on the subject include *Gaia: A New Look at Life on Earth* (1979), *The Revenge of Gaia* (2006), *The Vanishing Face of Gaia* (2009) and *A Rough Ride to the Future* (2014). In 2003 he was made a Companion of Honour and in 2005 *Prospect* magazine named him one of the world's top 100 public intellectuals. In 2006 he received the Wollaston Medal, the highest Award of the UK Geological Society.

Novacene

The Coming Age of Hyperintelligence

JAMES LOVELOCK

with Bryan Appleyard

PENGUIN BOOKS

PENGUIN BOOKS

UK | USA | Canada | Ireland | Australia
India | New Zealand | South Africa

Penguin Books is part of the Penguin Random House group of companies
whose addresses can be found at global.penguinrandomhouse.com

First published by Allen Lane 2019
Published in Penguin Books 2020
001

Copyright © James Lovelock and Bryan Appleyard, 2019

The moral right of the authors has been asserted

Typeset by Jouve (UK), Milton Keynes
Printed and bound in Great Britain by Clays Ltd, Elcograf S.p.A.

A CIP catalogue record for this book is available from the British Library

ISBN: 978–0–141–99079–8

Contents

CONTENTS

Preface

It is a great honour to have helped James Lovelock finish what will probably be his last book. I say 'probably' because experience has taught me never to guess what Jim will do next. Though he is now a very old man, a quiet retirement seems the least likely prospect, but he is toying with the idea, as he admitted in an email.

'Now that I am approaching 100 years it is easy to believe that I have little more to contribute. Like running a marathon, I know the agony of running that last hill that faces me. I might as well stop trying and let the young runners complete the course.'

I laughed when I read this; first, because I find it hard to imagine any young runner replacing Jim and, secondly, because I did not believe him. The truth is there could always be another book, just as there are always new ideas, new ways of looking, new ways of thinking. While working with him on this book, I actually had to ask him to stop thinking and start explaining, otherwise the task would never have been completed.

Jim's imagination is as thrillingly unexpected as it is alarmingly incisive. I once saw him sitting silently at a dinner party full of very bright, very serious people and then stun them into silence with a single sentence that overturned everything they had just been talking about.

And he always grows suspicious when he finds people agreeing with him – 'What have we got wrong?' he asks. He constantly looks for refutations and for different perspectives and he insists on the inherent uncertainty of scientific ideas. This makes his own ideas very robust indeed; they have been tested to destruction so many times. It is, of course, how all scientists should think and work, but many don't, which is why, in recent years, Jim has taken to calling himself an engineer.

He can be bewildering on first contact. I first met him many years ago at his laboratory in Coombe Mill. I just didn't understand him and I remember thinking I had fallen through a looking glass into a world utterly different from the one I thought I knew. He told me about his Gaia hypothesis, but I did not grasp the idea, perhaps because, as he says in this new book, it is not expressible in ordinary logical forms. This is not because it is complex – though, in detail, it is – but because at its core is a pristine simplicity. Life and the Earth are an interacting whole and the planet can be seen as a single organism; there you have it. Once I understood this it seemed so blindingly obvious that I assumed nobody could possibly disagree. In fact, back then, everybody did. Some still do, and some are Gaians but pretend they aren't, but most now acknowledge that Jim has altered forever our understanding of our lives and our planet.

People often talk of the value of 'thinking outside the box', but they seldom mention the much greater value of thinking, as Jim does, as if there is no box. He is so widely qualified – primarily in medical science and chemistry, but, seemingly, in everything once he starts talking – that no one discipline can ever hope to contain him. He is, as far as the institution of science is concerned, an outsider, a maverick, but that has not stopped him being garlanded with awards and honours. His nomination for the Fellowship of the Royal Society listed his work on the transmission of respiratory infections, air sterilization, blood-clotting, the freezing of living cells, artificial insemination, gas chromatography, and so on.

That was in 1974 and only briefly mentioned is the discipline that made him famous: climate science and his associated work on the possibility of extraterrestrial life. Then there is his ability to invent and construct his own gadgets – notably the revolutionary electron capture detector, maybe even the microwave oven and the numerous secret gadgets he created while working for the intelligence service.

Now, forty years after he introduced us to his goddess in his book *Gaia: A New Look at Life on Earth*, he introduces us to a new idea just as astounding and just as radical. 'Novacene' is Jim's name for a new geological epoch of the planet, an age that succeeds the Anthropocene, which began in 1712 and is already coming to a close. That age was defined by the ways in which humans had attained the ability to alter the geology and ecosystems of the entire planet. The Novacene – which Jim suggests may have already begun – is when our technology moves beyond our control, generating intelligences far greater and, crucially, much faster than our own. How this happens and what it means for us are the story of this book.

This is not the violent machine takeover seen in many science-fiction books and films. Rather, humans and machines will be united because both will be needed to sustain Gaia, the Earth as a living planet. As Jim put it to me in an email, 'The important concept, as I see it, is life itself. Perhaps this explains why I see the Earth as a form of life. The nature of its individual components, so long as they share a common purpose, seems unimportant.' Embodied in the concept of life is the possibility of knowledge, of creatures that can observe and reflect upon the nature of the cosmos. Whether humans continue living with or are superseded by their electronic progeny, we shall have played a vital and necessary part in the process of cosmic self-knowledge.

Jim is no anthropocentrist. He does not see humans as supreme beings, the summit and centre of creation. This was implicit in the idea of Gaia, which made it clear, to those who understood, that the biosphere has its own values of survival that lie far above and beyond any humanist values. It is explicit here: if life and knowing is to become entirely electronic, so be it; we played our part and newer, younger actors are already appearing on the stage.

Finally, a note on Jim's use of certain words. He uses 'cosmos' rather than 'universe' because he takes the former to mean everything we can know or see; he sees 'universe' as potentially meaning something larger of which we know and can know nothing. He uses 'cyborgs' to mean the intelligent electronic beings of the Novacene. In common usage this is taken to mean entities that are part flesh, part machine. But Jim thinks his usage is justified because his cyborgs will be products of Darwinian selection, and this they share with organic life. That will be all we share with the cyborgs; we may be their parents but they will not be our children.

Jim ended one of his recent emails with an apologetic, rhetorical sigh – 'That seems more than enough for now.' Enough for then, maybe, but not enough for James Lovelock for whom and from whom there will always be more.

Bryan Appleyard, 1 January 2019

PART ONE
The Knowing Cosmos

I

We Are Alone

Our cosmos is 13.8 billion years old. Our planet was formed 4.5 billion years ago and life began 3.7 billion years ago. Our species, *Homo sapiens*, is just over 300,000 years old. Copernicus, Kepler, Galileo and Newton appeared among us only in the last 500 years. For all but a brief moment of its existence the cosmos knew nothing of itself. Only when humanity developed the tools and the ideas to observe and analyse the bewildering spectacle of the clear night sky did the cosmos begin to awaken from its long sleep of ignorance.

Or did such an awakening also happen elsewhere? The inexhaustible flood of literature and films about aliens suggests we like to think so. It is difficult to believe we are alone in a cosmos which contains perhaps 2 trillion galaxies, each containing 100 billion stars. Some think that there is, surely, a chance that there have been or are highly intelligent species on at least one of the quadrillions of other planets that must orbit these stars. They would be, like us, understanders of the cosmos; or maybe their alien senses perceive an entirely different cosmos.

I think this is highly unlikely. These huge numbers of cosmic objects are misleading. It took the blindly groping process of evolution through natural selection 3.7 billion years – almost a third of the age of the cosmos – to evolve an understanding organism from the first primitive life forms. Furthermore, had the evolution of the solar system taken a billion years longer, there would be no one alive to talk about it. We would not have had time to attain the technological ability to cope with the increasing heat of the Sun. Seen from this perspective, it is clear that, ancient as it is, our cosmos is simply not old enough for the staggeringly improbable chain of events required to produce intelligent life to have occurred more than once. Our existence is a freakish one-off.

But our planet is now old. It is a curious fact that the lifespan of the Earth is easier to understand than our own lifespan. We do not yet know why humans rarely live beyond a maximum of 110 years and mice not much more than one year. It is not a matter of size – some small birds live to an age comparable with ours. In contrast, the lifespan of a planet is easily determined by the properties of the star that warms it.

Our star, the Sun, is what the astronomers call a main sequence star. It gave us life and it sustains us. Its warmth and regularity consoles us amidst the myriad uncertainties of our own lives. As that great truth-teller George Orwell wrote in 1946 in 'Some Thoughts on the Common Toad', 'The atom bombs are piling up in the factories, the police are prowling through the cities, the lies are streaming from the loudspeakers, but the earth is still going round the sun . . .'

But this great consoler is also lethal. Main sequence

stars slowly increase their brightness as they grow older. Increasing heat from the Sun threatens life on our planet. We have so far been protected by the planetary system I call Gaia which cools the Earth's surface.

There are several reasons why the Earth's temperature could become uninhabitably high. If there were no vegetation to absorb carbon dioxide (CO_2) it could not be lowered to its present levels. There would be a runaway greenhouse effect. We see evidence of this process around us all the time. If on a hot day you compare the temperature of a slate roof with that of a nearby black conifer tree, you would find the roof is 40 degrees hotter than the tree. The tree cools itself by evaporating water. Similarly, the sea surface is cool because life keeps it below 15°C; above that temperature there can be no sea life and sunlight is absorbed, heating the water.

Gaia must continue her work of cooling the planet, because it is now old and frail. With age, as I am all too well aware, we become more fragile. The same is true for Gaia. She could now be destroyed by shocks to her system which, in previous ages, she would have simply shrugged off.

I am pretty sure that only Earth has incubated a creature capable of knowing the cosmos. But I am equally sure that the existence of that creature is imperilled. We are unique, privileged beings and, for that reason, we should cherish every moment of our awareness. We should now be cherishing those moments even more because our supremacy as the prime understanders of the cosmos is rapidly coming to end.

2

The Edge of Extinction

I do not mean we're all going to die in the next few years – though it is possible. Human extinction has always been an imminent risk. We are very fragile understanders, precariously clinging to Earth, our only home.

Asteroid strikes could destroy the biosphere on which we depend, just as one seems to have ended the reign of dinosaurs 65 million years ago. The surfaces of the Moon and of our sister planet Mars are pockmarked with craters that almost certainly resulted from impacts with rocks. There is every reason to believe that the Earth has encountered just as many, but our planet, which has a thin, fluid skin of water, can show craters only on the land and these are washed away by the never-ceasing rain. Even so, if you examine the surface rocks carefully, as geologists have done, there is evidence of numerous impacts, some leaving craters as large as 200 miles in diameter.

Even more devastating would be a volcanic event like the one that – 252 million years ago – ended the Permian Period and started the Triassic. This, it is thought, was caused by a vast outflow of magma that formed what we now know as the Siberian Traps. The event is often referred to as the Great Dying – 90 per cent of marine species and 70 per cent of land organisms were extinguished. Ecosystems did not recover for 30 million years.

That was a long time ago, but is still no cause for complacency. Only 74,000 years ago the human population was massively reduced, perhaps to as low as a few thousand, by the volcanic winter that spread around the globe after the monstrous eruption that formed Lake Toba in Indonesia. And as recently as 1815, again in Indonesia, the eruption of Mount Tambora darkened the skies and lowered temperatures around the entire planet. This darkness is said to have inspired Mary Shelley's novel *Frankenstein* and Lord Byron's chilling poem 'Darkness', which ends: 'The winds were wither'd in the stagnant air,/And the clouds perish'd; Darkness had no need/ Of aid from them – She was the Universe.' The poet had glimpsed the cosmic fragility of our existence. Even if another such event did not wipe us out entirely, it could end our civilizations and send us back to the Stone Age. Understanding the cosmos would not then be high on our list of priorities.

Some of these risks can be mitigated. Thanks to our ability to understand, we already have rockets and nuclear weapons that could be used to deflect an asteroid that threatened the Earth. It should be a source of pride – though perhaps only temporarily – that we have so far been successful in not using those same weapons to destroy ourselves. If we have the international will to construct a rocket carrying a deflection package, then, for the first time ever, a planet of the solar system, the Earth, will have evolved the capacity to sense the approach of a large rock blundering through space on a deadly collision course and, much more than this, it will have evolved the means and the power to deflect its dangerous trajectory and save itself. In cosmic terms, this is a highly significant development.

Not all survival plans are quite as promising as that. One truly crazy idea for human survival appears at regular intervals in the media and in the minds of the venturesome. This is the notion that Mars could be a refuge for humanity if our life on Earth was in danger of being terminated. The assumption seems to be that the surface of Mars is not so different from that of the Saharan or Australian deserts. All that will be needed would be to drill down to an aquifer, just as they do in cities like Phoenix or Las Vegas in the United States. Then we could lead a comfortable civilized Martian life replete with casinos, golf courses and swimming pools.

Unfortunately, one thing the unmanned expeditions to Mars have told us is that the Martian desert is wholly inimical to all conceivable forms of Earth life. The atmosphere is about a hundred times thinner than the summit of Everest and it provides no shield against cosmic radiation or the ultraviolet radiation of the Sun. The thin air of Mars is 99 per cent CO_2 and utterly unbreathable. There are traces of water on the planet, but it is as salty as the waters of the Dead Sea and undrinkable. The pioneer and would-be spacefarer Elon Musk has said he would like to die on Mars, though not on impact. Martian conditions suggest death on impact might be preferable.

Perhaps Mars could provide hermit cells for the ultra-rich who might spend half their fortunes on voluntarily travelling there. Whatever cash was left could be spent on building and maintaining a tiny capsule of life from which escape would be impossible. It would actually be far less cruel to allow them to build their prison cells on the ice cap of Antarctica. At least the air is breathable.

To plan such ventures while ignoring the true state of the Earth seems extraordinarily perverse. The hope of finding some tiny Martian oasis does not really justify its enormous expense, especially when research costing a mere fraction of that of planetary exploration could provide crucial data about the Earth. We must never forget that this is the planet on which we live and that information about the Earth, although less exciting than news from Mars, may be the one thing that can ensure our survival.

So what do we need to know about the Earth to ensure that an understanding of the cosmos endures? We need to concentrate on heat, the most pressing and probable threat to our home and our existence.

I shall deal with this in more detail in the next part of this book, but I need to make a few points here. In recent years we have discovered thousands of 'exoplanets' – planets beyond our solar system. This has caused great excitement, not just among astronomers. Many have begun to speculate that we may be on the verge of finding signs of intelligent, organic alien life. But I suspect these people are being too anthropocentric. For one thing, it is important for alien-hunters to distinguish planets regulated by organic life forms from those regulated by electronic life.

That the latter will evolve from the former is the subject of this book. Any more advanced civilization than ours is likely to be electronic so there is little point in looking for small creatures with big heads and large, slanting eyes.

Then there is the matter of the temperature of these exoplanets. Particularly exciting has been the discovery that some lay within the 'zone of habitability'. This is sometimes called the Goldilocks Zone: like Goldilocks's porridge, it is just right – not too hot and not too cold. A Goldilocks planet would be just far enough from a star to support life – not so far as to become an ice world and not so near as to be sterilized by heat.

As I say, I don't think there are intelligent beings out there, but let's pretend for a moment that there are and they are doing exactly what we are doing – seeking planets in this habitable zone. These alien astronomers would reject Mercury and Venus, which are obviously too close to the Sun. But they would also reject Earth, which is also too close. Mars, they will conclude, is the only contender.

Earth absorbs and radiates such a prodigious amount of heat that it cannot possibly be classified as lying within the habitable zone. An alien astronomer viewing the solar system would be obliged to wonder about the anomalous surface temperature of our planet compared with that of Venus. The effective temperature of the Earth when seen from outer space is hotter, not cooler, than Venus. Yet the Earth is 30 per cent further from the Sun than Venus. The Earth's effective temperature is high because our atmosphere contains only a trace amount of carbon dioxide when compared with Venus. To stay in thermal equilibrium with the Sun, the Earth must radiate more thermal energy, and it does so at the long wavelengths of infrared. This makes the upper atmosphere at the edge of space hot, but, by the same measure, keeps the Earth's surface cool.

I think the zone of habitability idea is flawed because it ignores the possibility that a planet bearing life will tend to modify its environment and climate in a way that favours the life upon it, as ours does. A great deal of time may have been wasted during the search for life elsewhere because of the false assumption that the current environment of the Earth is simply a matter of geological happenstance. The truth is that the Earth's environment has been massively adapted to sustain habitability. It is *life* that has controlled the heat from the Sun. If you wiped out life entirely from the Earth, it would be impossible to inhabit because it would become far too hot.

So we are made by our star, which provides the energy for life, but we are also threatened by it. This star is a perfectly ordinary, somewhat small, middle-aged cosmic entity – a 5 billion-year-old main sequence star. Models of the Sun explain how it stays hot by fusing its hydrogen into helium in the ultra-incandescent regions of its interior. But just as burning coal in oxygen produces carbon dioxide, so fusing hydrogen produces helium. Both carbon dioxide and helium are greenhouse gases: the first warms the Earth, the second warms the Sun. This makes the inner regions of the Sun hotter and so increases the rate of fusion; the extra heat makes the Sun expand and from its greater surface area more heat escapes and warms the Earth. It will continue to increase its output of heat until, in 5 billion years' time, it becomes a red giant star and slowly absorbs the Earth and the inner planets of the solar system.

So far, the heating of the Sun has been slow enough to allow life to evolve, a process which takes millions of years. Unfortunately, the Sun is now too hot for the further development of organic life on Earth. The output of heat from our star is too great for life to start again as it did from the simple chemicals of the Archean Period between 4 billion and 2.5 billion years ago. If life on Earth is wiped out, it will not start again.

But that is not the immediate problem. The real threat is that, even though for the moment it is stable, the Sun is gradually emitting ever more heat. In fact, over the last 3.5 billion years its output has increased by 20 per cent. This should have been enough to raise the surface temperature of the Earth to 50°C and bring about a runaway greenhouse effect that would have sterilized the planet. But it didn't happen. To be sure, there have been what we feel to be hot periods and ice ages, but the average temperature of the whole planetary surface does not seem to have varied by more than about 5°C from its current temperature: 15°C.

Gaia does this. In Greek mythology, Gaia is the Greek goddess of Earth and, at the suggestion of the novelist William Golding, I gave her name to the theory I developed fifty years ago. The theory is that, since it began, life has worked to modify its environment. This is not easily explained in full because it is a complex, multi-dimensional process. I can, however, illustrate how it works with a simple computer simulation. This is called Daisyworld, which, with the atmospheric scientist Andrew Watson, I published in 1983.

A main sequence star like our Sun gradually heats the planet Daisyworld until it is just warm enough for a species of black daisies to colonize the entire surface. Black daisies absorb heat so they thrive in these low temperatures. But there are mutant white daisies which reflect heat and, as the temperature rises even further, these begin to flourish. So Daisyworld is cooled by white daisies and warmed by black ones. A simple flower is able to regulate and stabilize the environment on a planetary scale. Moreover, this stabilization emerges from a strictly Darwinian process.

Scale up this model to include all the flora and fauna of Earth and you have the system I have called Gaia. In fact, you cannot actually scale it up because the system is too complex; so complex in fact that we are nowhere near fully understanding it. Perhaps it is hard to understand because we are an intrinsic part of it. But also, I suspect, it is because we have been too reliant on language and logical thinking and have not paid enough attention to the intuitive thinking that plays such a large part in our understanding of the world.

So, in short, humans may, at any moment, become extinct because of forces far beyond our control. But we can do something to save ourselves by learning to think.

3

Learning to Think

Gaia is not easy to explain because it is a concept that arises by intuition from internally held and mostly unconscious information. This is quite different from the concepts that arise directly from the stepwise logic preferred by scientists. Dynamic, self-regulating systems wholly defy a logical explanation that uses step-by-step arguments. So I cannot give you a logical explanation of Gaia. Nevertheless, to me, the evidence for her existence is very strong indeed. You will find it outlined in detail in my books and papers.

I have often been criticized for the suggestion – which seems to me to be intuitively true – that Gaia shows that the entire Earth is a single living organism. One argument against this is that it cannot be a living organism because it cannot reproduce. My response to that is that no 4-billion-year-old organism needs to reproduce. Perhaps I would also say that if those non-existent aliens saw one of those anti-asteroid rockets emerging from Earth's atmosphere, they might reasonably conclude it had been launched by the planet itself. They would be right, precisely because it is the entire system – Gaia – which has produced that rocket. But they would be wrong to think the proximity to a star and the heat radiating from Earth indicated that life could not happen here. That radiation is the work of Gaia. It is she who pumps excess heat out into space to preserve life and it is for her sake that we must change our ways of thinking.

As a much younger man I accepted the conventional scientific view that the cosmos was a straightforward system of cause and effect. B is caused by A and then causes C. I had perhaps not paid close enough attention to Gaia. The 'A causes B' way of thinking is one-dimensional and linear whereas reality is multi-dimensional and non-linear. One has only to think of one's own life to see how absurd it is to think everything can be explained as a simple linear process of cause and effect.

There are also examples from basic engineering. Take the steam engine governor invented by James Watt in the nineteenth century. This was a solution to controlling the speed of a locomotive. The governor consists of a vertical steel shaft rotated by a tiny proportion of the main drive power and it simply spins out a pair of brass balls. The faster the rotation, the more they spread out. The movement of the spinning was arranged so that fast spinning closes the valve that sets the quantity of steam passing to the engine. For any given setting, this simple system would stabilize and maintain a constant speed regardless of whether the engine was going uphill or downhill. Using it, the driver could set a constant speed and leave the governor to maintain it.

You might think that this is simple and obvious; clever, but no more than that. Think again. Trying to explain how the governor worked was beyond the power of the greatest physicist of the nineteenth century, James Clerk Maxwell. He reported to the Royal Society that he lay awake for three nights trying to explain how it worked, but he failed.

So the pure and concise linear logic that we trace back to Aristotle – logic which is the basis of so much that is important in science and in human affairs – utterly fails to explain simple systems like the steam engine governor. Much more than this, the temperature regulation of an animal or Gaia is equally inexplicable by classical logic.

The mistake I think we made was to continue to reason classically. We made this mistake because of the nature of speech, either spoken or written, and the fissiparous tendency of human thinking. We know that our friends and lovers are whole persons. It may seem sensible at various times to consider their livers, skin and blood to understand their special function, or for purposes of medicine, but the person we know is much more than the mere sum of these parts.

As I see it, the logical problem with speech is that it proceeds step by step linearly. This is fine for the solution of essentially static problems and has served us well; it has led logicians such as Frege, Russell, Wittgenstein and Popper to offer comprehensible explanations of our world.

Now, when I look back on the long arguments I had over Gaia with evolutionary biologists in the Western world, they seem to have been arguments at cross purposes. From the beginning I saw Gaia as a dynamic system.

I knew instinctively that such systems cannot be explained in linear logical terms, but I did not know why. My instinct arose because I was intimately familiar with scientific instruments that operated dynamically. Equally importantly, I had started work in 1941 at the Department of Physiology of the National Institute for Medical Research. Here the scientists were system scientists. My young mind took for granted the non-linear way of thinking of dynamic systems.

Whilst it is true that the Gaia hypothesis was unacceptable to many Earth and life scientists in the English-speaking world, European scientists tended to be more open-minded. The eminent Swedish scientist Bert Bolin, and other members of the European Geophysical Union, made sure that peer reviewers did not obstruct the publication of my first detailed paper on the Gaia hypothesis in 1972 in the Swedish journal *Tellus*. More recently, the outstanding French savant Bruno Latour has given support to Gaia and sees it as the natural successor to Galileo's vision of the solar system as a collection of planets made of rock and rotating around the Sun. In Latour's vision it is the similarity of the planets that is significant. In the Gaian version it is the extraordinary difference of the Earth from the other planets that makes it special.

With few exceptions, the battle over Gaia was a gentlemanly affair. Unusually for a scientific contest, we had agreed to disagree. It is important to note that, had I been dependent on grants to fund my research, it could not have been done. In practice, all the costs, including my income and travel, came from payments I received for solving the technical problems of government services and industry. Academia almost everywhere behaved, albeit more gently, like the Church in Galileo's time. I find it extraordinary that so many good scientists should have befuddled themselves trying to force classical logic to explain the inexplicable. But then an even larger number of clerics have done the same.

As Newton found long ago, logical thinking does not work with dynamic systems, things that change over the course of time. Quite simply, you cannot explain the working of something alive by cause-and-effect logic. Most of us, especially women, have known this all along.

Newton made his discoveries in the seventeenth century while working in an environment soaked in classical thinking – Trinity College, Cambridge. Wisely, he disguised and transformed the logic of dynamic systems into something he called calculus. Ever since then, other mathematically inclined scientists have paid homage to Newton and have invented methods that apparently allow otherwise inexplicable dynamic systems to be analysed.

I think of these physicists now engaged in the wonders of quantum computers and other practical applications of quantum theory as akin to engineers and physiologists. Have they recognized intuitively that, although the wonders they produce and invent are real and work, they can never explain them? The most they can do is to describe them.

I also wonder if those great minds possessed by Newton, Galileo, Laplace, Fourier, Poincaré, Planck and many others thought intuitively in a way similar to that of the cathedral builders. None of them possessed even slide-rules to calculate that subtle balance of a beautiful column so that it was strong enough to last for centuries. Next time you drive across a mile-long suspension bridge or fly at 40,000 feet, remember that the mathematics used to design the bridge or the plane came from something quite illogical. What the engineers did was to use an honourable deceit. It seemed to explain how the system worked, but really it did no more than describe it.

I have applied this same honourable deceit to make usable the illogical mathematics of ecosystems, but so far almost no one has used it. In 1992 I published a paper in the *Philosophical Transactions of the Royal Society* that was based on a conjecture by the biophysicist Alfred Lotka. He proposed that, contrary to expectation, it would be easier to model an ecosystem of many species if the physical environment were included, a very Gaian conclusion.

Before speech and writing appeared, we and all other animals thought intuitively. Imagine a country walk where you come unexpectedly to the edge of a cliff, so high and so steep that one further step would lead to certain death. If this happens, your brain analyses the vision before you and in milliseconds unconsciously recognizes the danger. All further forward motion is inhibited. Recent measurements show that this instinctive reaction operates within 40 milliseconds of the recognition of danger. It happens well before you are conscious of the cliff. In other words, you are saved by instinct, not by rational conscious thoughts about the danger of falling. Human civilization took a bad turn when it began to denigrate intuition. Without it, we die. As Einstein said, 'The intuitive mind is a sacred gift and the rational mind is a faithful servant. We have created a society that honours the servant and has forgotten the gift.'

Perhaps it happened because women's insights were rejected. How long ago was it when the first group of men referred to an idea they didn't like as 'mere feminine intuition'? I suspect it occurred when we moved from hunting and gathering to living in cities. It was certainly embedded in ancient Greek philosophy. Socrates' remark, 'Nothing interesting happens outside the city walls', seems applicable to urban life. But it comes at the cost of valuing conscious thought, debate and argument more highly than instinct. Conscious debate cost Socrates his life.

So, the unconscious mind can perceive danger within 40 milliseconds, a time too short for conscious awareness. More than this, within that fragment of unconscious thought there was time for the intuitive part of the mind to come up with a muscular response. This is how we evolved to escape the faster and more powerful predators.

Science is never certain or exact. The best we can ever do is to express our knowledge in terms of its probability. We must understand that we are still primitive animals. There is a vast amount to be discovered about our universe that is comprehensible, but an unknown and probably far greater amount of it is quite ineffable and, as today, we will never understand it all.

Because of a passionate desire for certainty, something

that may have evolved during our hunter-gatherer phase, the information we have gathered about our world and the universe may be tinted by the colour of our faith or, more recently, political belief, but I think that this matters very little because as we grow wiser it is easy to distinguish the gem from the mud that encases it.

Few things illustrate more clearly how misleading cause-and-effect logic can be than the invention of the non-existent planet Vulcan. Observations in the early nineteenth century of planet Mercury's orbit revealed it to be anomalous compared with the orbits of the other planets of the solar system. If the orbital deviation was accepted as true, it meant that Newton's laws of planetary motion were in error. Rather than accept so disturbing a possibility, scientists invented the planet Vulcan, put it in an imaginary orbit closer to the Sun and gave it a mass whose gravitational attraction was sufficient to account for the deviation of Mercury's orbit.

Nearly a century later Einstein proposed that the deviation of Mercury's orbit was a consequence of the relativistic distortion of space-time by the huge mass of the Sun. Astronomers still accepted Newton's laws but recognized that they failed to provide a complete answer in regions of high gravity.

This is an example of a planet-sized mistake that can occur if cause-and-effect logic is taken literally. It was thinking about Vulcan that made me look out to the western sea horizon while walking along the Dorset coast just after the Sun had set. The sky was darkening, now that the Sun was below the sharp line of the western horizon, and I was rewarded by the sight of Mercury twinkling close to the horizon. It is a rare enough sight here at 52 degrees North and I wondered whether, had the hypothetical planet Vulcan been real, anyone would have seen it, or would it have always been hidden in the Sun's effulgence? We are where we are and we see only what can be seen. But, with intuition, we can know now far more than we can see.

4

Why We Are Here

In Douglas Adams's *Hitchhiker's Guide to the Galaxy* books dolphins are smart enough to leave the Earth just before it is destroyed. Their departing message to humanity is, 'So long, and thanks for all the fish.' Like all the best jokes, this works because it gives us the uneasy feeling that it may not be just a joke. We know whales, octopuses and chimpanzees are clever, but what do they think about? How do they use their intelligence? Maybe, like the dolphins, they see us as a messy, rather stupid species, useful primarily for the provision of food.

Adams dramatizes this feeling by making his dolphins conscious of an imminent threat to Earth and capable of leaving. I would not rate dolphin intelligence quite so highly. To me it is clear that, however intelligent other creatures may be, the distinguishing feature of human intelligence is that we use it to analyse and speculate about the world and the cosmos and, in the Anthropocene, to make changes of planetary significance. As I have said, I believe only we do this, only we are the way in which the cosmos has awoken to self-knowledge.

So not only would human extinction be bad news for humans, it would also be bad news for the cosmos.

Assuming I am right and there are no intelligent aliens, then the end of life on Earth would mean the end of all knowing and understanding. The knowing cosmos would die.

At this point, I need to go back to my student days in the 1930s. Then, it was quite normal for most people in Britain to say they believed in God. In those earlier times religion was much more a part of life and many believed that God had made humans special chosen people. Now that God no longer has supreme status, do we still see ourselves as chosen?

Probably not, but I do. Perhaps because I was brought up as a Quaker, I do not have a literal view of religion – I accept much of its wisdom but not necessarily the truth of the stories. I now think that this religious view of humanity as chosen may express a deep truth about the cosmos. This thought was first inspired by a book published in 1986 entitled *The Anthropic Cosmological Principle* by two cosmologists, John Barrow and Frank Tipler.

The first effect of Barrow and Tipler's book was to throw up fireworks of doubt in my mind about the scientific principle of cause and effect. Latterly I have realized that I have never really been a pure scientist, I have been an engineer. All the instruments I have invented are based on engineering principles (although I often built them with a conviction that they are possible because I have been able to prove it scientifically). Engineers start from the world as it is rather than from a scientific principle. This happened when I received a letter from NASA in 1961. They invited me 'to participate in the first Surveyor mission . . . which is scheduled for a lunar soft landing' two years later. They wanted me to help make a gas chromatograph. It had to be as small as possible. I knew at once I could do it, even though I did not know how.

The second effect of the book was to make me think we may indeed be chosen. Barrow and Tipler start from the anthropic principle. This sounds like a purely philosophical argument, but in fact it has serious implications for science. In its most basic form it says something which may, on reflection, seem obvious. This is that in attempting to describe the cosmos we must first assume that it must be the kind of cosmos that can produce thinking beings like us. In other words, we can't come up with a theory in which the cosmos is too young or consists entirely of radiation or one in which the Earth never came into existence. Our theories are limited by the fact that we are here to dream them up.

So anything we say about the cosmos cannot, if we aspire to truth, disprove the existence of thinking creatures capable of saying those things. We know, for example, the cosmos must be more than, say, a million years old because it would take much longer than that for intelligent beings to evolve. This means our very existence restricts what we can say about the cosmos. This is controversial because some think it is a banal statement that adds nothing to our knowledge. I disagree.

Barrow and Tipler go further than this. When we do observe the cosmos we find that it appears to be fine-tuned to produce us. There are many physical constants, any one of which could have been even fractionally different and we would not have existed. Perhaps we are staggeringly lucky – the products of a mass of extraordinary coincidences – but this does not constitute any kind of explanation.

One response to this is to say that God must have been responsible for the benign conditions; how else to explain something which evades all scientific explanation? Or, secondly, we could argue, as many do, that there are multiple universes and, obviously, we are in the one in which intelligent life could appear; there would be nothing magical about that. This 'multiverse' theory is used as one explanation of the mysteries of quantum theory. It is no surprise that one cosmos out of billions has been set up to produce life; others just carry on, unknowing and unknown. That seems to me to be no more than a get-out-of-jail-free card – because it can be neither proved nor disproved.

But Barrow and Tipler provide a third option. Perhaps information is an innate property of the universe and, therefore, conscious beings must come into existence. We would then really be the chosen people – the tool whereby the cosmos would explain itself.

So can we say the purpose of the cosmos is to produce and sustain intelligent life? This is tantamount to a religious statement – not in the sense of the stories in which I don't believe but in the sense of a deep truth in which I do. That outstanding leader of Czechoslovakia and then the Czech Republic, Václav Havel, stated in 2003 when he received the Freedom Prize in Philadelphia that the cosmic anthropic principle and Gaia were two hypotheses that pointed a seemly way to the future. The connection was right and profoundly true.

I find it deeply moving to consider how, from its origin at the Big Bang, our universe was formed – first, the light elements from which the early stars and galaxies were born; then, over billions of years, the elements of life slowly accumulated, as did the star systems on whose planets they could combine and eventually form the first living cells. Then, over another 4 billion years, chance and necessity led to the evolution of animals and, eventually, humans. Could it have happened differently? No, according to Barrow and Tipler. And we may only be the beginning, the start of a process whereby the entire cosmos attains consciousness.

Where the new atheists and their secular fellow travellers have gone wrong, I think, is that they have thrown the baby of truth out with the bathwater of myth. In their dislike of religion, they have failed to see its inner core of truth. I think we are a chosen people, but not chosen directly by God or some individual agency; instead, we are a species that was selected naturally – selected for intelligence.

At this point we risk being drawn into the quasi-theological discussions surrounding quantum theory, the world of the very small whose mysteries inspire many competing explanations. It may be that the cosmic anthropological principle put forward by Barrow and Tipler is the most sophisticated religious concept yet devised. But we don't need to go that far to embrace the idea that we are, indeed, chosen. This gives us cause to be proud, but not to be full of hubris, for our status carries enormous responsibilities. Think of us as a species like the first photosynthesizers. Those primitive unicellular organisms unconsciously discovered how to use the flood of energy in sunlight to make the food needed for their progeny and, at the same time, release into their world that magic – though deadly for many organisms – gas, oxygen. Without them there would now be no life on this planet. I think that our emergence as a species is as important as was the emergence of those light-harvesters 3 billion years ago.

It is a cause for pride and joy that we can harvest sunlight and use its energy to capture and store information, which is also, as I shall explain later, a fundamental property of the universe. But it demands that we use the gift wisely. We must ensure the continued evolution of all life on Earth so that we can face the ever-increasing hazards that inevitably threaten us and Gaia, the great system comprising all life and the material parts of our planet.

We alone, among all the species that have benefited from the flood of energy from the Sun, are the ones who evolved with the ability to transmute the flood of photons into bits of information gathered in a way that empowers evolution. Our reward is the opportunity to understand something of the universe and ourselves.

5

The New Knowers

But, as I said earlier, our reign as sole understanders of the cosmos is rapidly coming to an end. We should not be afraid of this. The revolution that has just begun may be understood as a continuation of the process whereby the Earth nurtures the understanders, the beings that will lead the cosmos to self-knowledge. What is revolutionary about this moment is that the understanders of the future will not be humans but what I choose to call 'cyborgs' that will have designed and built themselves from the artificial intelligence systems we have already constructed. These will soon become thousands then millions of times more intelligent than us.

The term 'cyborg' was coined by Manfred Clynes and Nathan Kline in 1960. It refers to a cybernetic organism: an organism as self-sufficient as one of us but made of engineered materials. I like this word and definition because it could apply to anything ranging in size from a micro-organism to a pachyderm, from a microchip to an omnibus. It is now commonly taken to mean an entity that is part flesh, part machine. I use it here to emphasize that the new intelligent beings will have arisen, like us, from Darwinian evolution. They will not, at first, be separate from us; indeed, they will be our offspring because the systems we made turned out to be their precursors.

We need not be afraid because, initially at least, these inorganic beings will need us and the whole organic world to continue to regulate the climate, keeping Earth cool to fend off the heat of the Sun and safeguard us from the worst effects of future catastrophes. We shall not descend into the kind of war between humans and machines that is so often described in science fiction because we need each other. Gaia will keep the peace.

This is the age I call the Novacene. I'm sure that one day a more appropriate name will be chosen, something more imaginative, but for now I'm using 'Novacene' to describe what could be one of the most crucial periods in the history of our planet and perhaps even of the cosmos.

Before exploring the Novacene, I need to describe how we reached this point through the workings of the age that preceded it. This is the period in which humans, the chosen species, developed technology which enabled them to intervene directly in the processes and structures of the entire planet. It is the age of fire in which we learned to exploit the captured sunlight of the distant past. It is known as the Anthropocene.

PART TWO

The Age of Fire

6

Thomas Newcomen

Born in Dartmouth, Devon, in 1663, Newcomen died in London in 1729. His post-mortem in the *Monthly Chronicle* noted that he was the 'sole inventor of that surprising machine for raising water by fire'. And so he was, but 'surprising' takes English understatement a little too far; 'world-transforming' would have been more accurate.

Little is known of the life of Thomas Newcomen. He was a Baptist lay preacher, a blacksmith and an engineer, though he was not educated as such. There is a story that he corresponded with the scientist Robert Hooke, but this may not be true. Not that he would have needed Hooke's help. He was a practical, not a theoretical man and he had a very practical problem to solve: he needed to find a way to get more coal out of the ground.

In Europe in the late seventeenth and early eighteenth centuries population increases, the formation of nation states and the ensuing wars had led to ever more demand for raw materials, notably wood. This was used on a massive scale for shipbuilding – at the beginning of the eighteenth century the building of a warship could consume 4,000 trees – and for smelting iron. Forests were being depleted faster than they could regrow. As a fuel, coal – which produces ten times more heat than wood – was the obvious replacement. But production was limited by flooding in the mines. For Britain, then rising to its ascendancy as a global superpower, this was an urgent problem.

Enter the engineer to transform, first, global strategy and then the globe itself. All Newcomen did was build a steam-powered pump. It burned coal and used the heat produced to boil water into steam which was let into a cylinder with a movable piston. The piston rose and then cold water from a stream nearby was sprayed into the cylinder; the steam condensed, the pressure dropped and the piston moved back to its starting position, doing a substantial amount of work in the process and clearing the mines of water. This 'atmospheric steam engine' was not the first steam engine, but it was the best so far and its descendants powered the railway engines of the nine-teenth century. For my purposes, what it was used for is of less significance than its impact.

This little engine did nothing less than unleash the Industrial Revolution. This was the first time that any form of life on Earth had purposefully used the energy of sunlight to deliver accessible work and do so in a way that was profitable. This ensured its growth and reproduction. It might be said that windmills and sailing ships did the same thing when they used wind to drive them. What was special about the appearance of Newcomen's engine was that it could be used anywhere at any time and it was not dependent on any vagaries of the weather. It spread across the world. I think that Newcomen's invention should be heralded not just as the start of the Industrial Revolution but also as the beginning of the Anthropocene – the age of fire, the age in which humans acquired the power to transform the physical world on a massive scale.

Humans had made machines before, but this was a wholly new type. Newcomen's machines could run in the absence of a human operator. This was not entirely original; clocks, for example, had run on their own since the first clepsydrae (water clocks), thought to date from 6,000 years ago. Newcomen's engine, however, was much more powerful and it could cause large-scale changes to the physical world. It was first used in a mine in Griff, a hamlet just south of Dudley in Warwickshire. By 1733, four years after Newcomen's death, around 125 of his engines had been installed in most of the important mining districts of Britain and Europe.

Newcomen had simply made coal, and with it energy, more easily accessible. His steam pump enabled the exploitation of a hitherto inaccessible fossil fuel. Prior to this the energy available to our species was, basically, the sunlight that fell on the surface of the Earth. This included the energy locked up in the trees and plants. Over millions of years plant material had become coal. Trees more than 200 million years earlier had captured solar energy and turned it into potential chemical energy in the form of oxygen and wood. The wood, when fossilized, became coal. Burning it releases concentrated ancient stores of solar energy, millions of years of sunlight captured in black stone.

At this point I want to stress that the evolution of the

Anthropocene – which has so massively changed the Earth – was driven by market forces. Had there been no economic gain from using Newcomen's steam engine, we might still be back in the world of the seventeenth century. The important feature of Newcomen's engine was its profitability. The mere idea of the engine would not have been enough to ensure its development. Its great significance – for good or ill – was because it was cheaper as a source of work than human or horse power.

7

A New Age

This was the tipping point, the start of a new age. In due course, it caused seismic social upheaval. The Industrial Revolution was an era that simultaneously brought great wealth and great poverty. It brought poverty because working men who previously could feed themselves and their family from the proceeds of selling their labour were impoverished by this new cheap source of labour. It brought wealth because these new artificial workers could produce far more than humans.

Though the term 'Industrial Revolution' is accurate enough, it neither catches the wider significance of the moment, nor does it encompass its full duration. The better name is the Anthropocene because it covers the full 300 years from Newcomen's installation of his steam pump until now and it captures the great theme of the era: the domination of human power over the entirety of the planet.

The word 'Anthropocene' was first used in the early 1980s by Eugene Stoermer, an ecologist who worked on the waters of the Great Lakes that separate Canada from the United States. He coined it to describe the effect of industrial pollution on the wildlife of the lakes. It was one more sign that, in the Anthropocene, human activity could have global effects.

My own contribution to this insight came in 1973. In the late 1950s I had invented – using non-linear, intuitive insight – a device called the electron capture detector (ECD). This worked by transforming the linear DC signal of the device into frequency, where the amount of substance detected was expressed directly as a frequency. The ECD is capable of detecting almost infinitesimally small amounts of chemical compounds. In 1971 I took one of these on a voyage to the South Atlantic and found traces of chlorofluorocarbons (CFCs) in the atmosphere. These were widely used in, among other things, fridges. Manufacturers were determined to deny they had any effects on the global environment, notably the depletion of the ozone layer in the atmosphere. My findings showed that CFCs had spread around the globe. They were first regulated and later banned.

Analytical chemistry provided evidence that we had entered a world where human inventions or innovations could affect the entire planet, the world of the Anthropocene. There are arguments about when this epoch began. Some put it as long ago as the first appearance of *Homo sapiens*, others as recently as the first atomic explosion in 1945. For the moment, it is not even generally accepted as a geological epoch. Many insist we are still in the Holocene, which began about 11,500 years ago when the last ice age ended. Before that was the Pleistocene, which lasted 2.4 million years, and before that was the Pliocene (2.7 million years) and the Miocene (18 million years). The numbers seem to rise consistently until you get back to the Big Bang, when they suddenly become impossibly small: the first age of the cosmos, known as the Grand Unification epoch, began 10^{-43} seconds after the Big Bang and only lasted until 10^{-36} seconds. If we accept the Anthropocene, as I believe we should, the ages are getting shorter again. In my view, the Novacene may only last 100 years, but I shall come back to that.

For me, the key point that justifies the definition of the Anthropocene as a new geological period is the radical change that took place when humans first began to convert stored solar energy into useful work. This makes the Anthropocene the second stage in the planet's processing of the power of the Sun. In the first stage the chemical process of photosynthesis enabled organisms to convert light into chemical energy. The third stage will be the Novacene, when solar energy is converted into information.

But if you really need more evidence that the Anthropocene is truly a new age – first, look around you at the spreading cities, the roads, the glass towers of offices and flats, the power stations, the cars and trucks, the factories and the airports. Or look at pictures of the Earth at night seen from space – it has become a quilt of shining points of light. Secondly, you should read *The Natural History of Selborne* by Gilbert White to see how far we have come. White was curate of the church in the village of Selborne, Hampshire. He was an observer and writer of genius – he wrote of the sound of a swallow catching a fly in its beak as 'resembling the noise at the shutting of a watch case'. This book, published in 1789, before the power of the

Anthropocene had become evident, is essential reading for anyone who wants to know how things were before the modern, fast-changing world of the new age became the norm. White was a polymath and a scientist. Like me, he made his own instruments and used them to make accurate observations of the natural environment.

His book is both a loving account of the natural world and a scientific text that is still useful today. He noted, for example, the inclement heat, cold and fog of the year 1783. It had been caused by the eruption of the Laki volcano in Iceland which emitted huge quantities of ash and sulphur gases that reacted with the air to form aerosols of sulphuric acid. Climatologists can now test the reliability of their experimental predictions by using the Laki eruption as if it were an experimental perturbation to see how well they agree with the climate change at Selborne.

From White's Selborne to the megacities of today with populations of 30 million or more is not simply a development, it is an explosive transformation of the world, a massive increase in the intensity of life on Earth. Nothing like it has ever happened before. The Anthropocene may not have been officially recognized; it is, nevertheless, the most important period in our old planet's long history.

8

Acceleration

Gilbert White's book was later seen as portraying a world we have lost and now lament. White was born in 1720, eight years after Newcomen installed his first steam pump, and he died in 1793, as the Anthropocene was closing in on the world he had celebrated and recorded. Then, in 1825, the new age fully took hold with the opening of the Stockton and Darlington Railway; thereafter, railways spread rapidly around the world. The story of the Anthropocene in the nineteenth century was about this global development. In China, now the world's primary industrial economy, the first railway was built in 1876 and, by 1911, there were 9,000 kilometres of track.

The advent of the railways introduced another of the great themes of the Anthropocene – acceleration. Soon after the Anthropocene began, we became like the boy racer, carried away by the power of acceleration. We have kept our foot on the accelerator for 300 years and are now approaching the time when our electronic, mechanical and biological artefacts can run the Earth system by themselves.

Previous technologies had not affected the speed of human movement; Napoleon's armies moved not much faster than Julius Caesar's. From the moment of the invention of trains, their speed steadily increased until they attained the 200 mph of today and the 400 mph of the Maglev trains of the future. Not only that, they moved large numbers of people who had previously relied on their feet or, if they were rich, horses. Imagine a railway line being constructed near a village deep in the countryside. Centuries of local experience and wisdom about how the world and your own life worked would be overturned at the sight of the first locomotive.

The Romantic poet William Wordsworth saw more clearly and painfully than most what was happening. His sonnet 'On the Projected Kendal and Windermere Railway' begins:

> Is then no nook of English ground secure
> From rash assault? Schemes of retirement sown
> In youth, and 'mid the busy world kept pure
> As when their earliest flowers of hope were blown,
> Must perish;– how can they this blight endure?

The Anthropocene makes no allowances, even for the longings of literary genius.

Beyond trains, Anthropocene acceleration has progressed far beyond anything Wordsworth could have imagined in his most fervid nightmares. Military aircraft can now travel at more than twice the speed of sound and rockets reach 25,000 mph, the speed needed to escape from the Earth's gravitational field. But the most world-changing speed is that of civilian aircraft – 500–600 mph – because they transport vast numbers of people around the world, extending the cultural homogenization and global reach of the new age.

Such developments signal another form of acceleration. The Anthropocene brought with it a new means of rapid evolution. The seabird, with its graceful flight, took more than 50 million years to evolve from its lizard ancestor. Compare this with the evolution of today's airliners from the string-bag biplanes that flew a mere 100 years ago. Such intelligent, intentional selection appears to be a million times faster than natural selection. By moving beyond natural selection, we have already enrolled as sorcerers' apprentices.

But the form of acceleration that has the most significance for the age that is now being born is electronic. In 1965 Gordon Moore, co-founder of silicon-chip maker Intel, published a famous paper in which he predicted that every year there would be a doubling in the number of transistors that could be installed on an integrated circuit. Known as Moore's Law, this means that the processing speed and capacity of silicon chips would increase exponentially.

With some variations – the doubling rate has been revised to two or slightly more years – Moore has been proved right and the doubling he predicted has continued for at least forty years. If you think doubling every second year is not that fast, then think again, because it means a thousand-fold increase in twenty years and a trillion-fold growth in a lifetime of eighty years. Some say this process will stall when we reach the physical limit of silicon. This may be true, but in the future chips are likely to be carbon-based; a diamond chip would have speed beyond anything we can now envisage.

9
War

Sadly, the power of the Anthropocene has manifested itself most forcibly in war. It has been an age made for increasingly bloody conflict, thanks to the ingenuity of our new machines. As the philosopher and historian Lewis Mumford noted in *Technics and Civilization*, 'War is the supreme drama of a completely mechanized society.'

Before 1700 warfare was brutal enough, but it was powered primarily by the hands of men and a certain amount of gunpowder. The American Civil War, which lasted from 1861 to 1865 and cost over a million lives, was the first occasion when war was powered also by the products of the Anthropocene. Richard Gatling's rapid-fire 'rotary cannon', the forerunner of all future machine guns, was first used in this conflict. Trench warfare also came of age, a product of the increasing speed and range of weaponry. It was a tactic that led to the mud-bound slaughter of the First World War.

Then came air power. This extended the battle lines to cover entire nations and civilians became legitimate targets. The ferocious bombing of Guernica in April 1937 by Hitler's Luftwaffe in support of Franco's fascists made it clear that, in the wars of the Anthropocene, there was no such thing as a non-combatant.

This entire grim history would have been even worse had Leo Szilard crossed a certain road in London a decade earlier than he did. Szilard was a Hungarian Jewish physicist who moved to London when Hitler rose to power in 1933. On the morning of 12 September of that year he stepped off the kerb to cross Southampton Row. As he did so, according to the historian Richard Rhodes, 'Time cracked open before him and he saw a way to the future . . . the shape of things to come'. He had seen the possibility of creating a nuclear chain reaction; he had seen nuclear power, but he had also seen the atom bomb. If that had been 1923 rather than 1933, the Second World War would have been fought with nuclear weapons. It would have been much shorter and even deadlier.

In the event, it was another twelve years before nuclear weapons could be manufactured. Only Hiroshima and Nagasaki suffered nuclear attacks. Subsequent nuclear explosions were all tests; these reached a terrible climax in 1961 with the Soviet Tsar Bomba, a 50-megaton fusion weapon which, if ever used in anger, would obliterate a large city and its immediate surroundings. The pollution from these tests was so huge that even today, nearly sixty years later, the resulting radioactivity of our bodies is useful to forensic scientists in establishing times of death.

The arms race in powerful nuclear weapons reached an absurdly dangerous state in the year of the Tsar Bomba. During that poisonous period, about 500 megatons of nuclear explosive were detonated on islands in the Pacific and Arctic oceans. This is a quantity equivalent to the unshielded explosion of 30,000 Hiroshima-sized atom bombs. This was insanity.

I shall never forget standing beside the warhead of a nuclear missile that had been opened for inspection. The three bombs it held were brightly coated with aluminium foil and each was small enough to fit in the hand. They were constructed so that any one of them could wreck a city the size of London. Each one was sixty times more powerful than the one that burst over Hiroshima more than seventy years ago. What politician or war leader would dare to launch one? For now, it would seem to be the ultimate crime.

We can take comfort from the fact that the use of such weapons has not been repeated in warfare in the seventy-plus years that have passed since then. It may be that their existence has been enough to deter great wars and their sheer deadliness has had the benign effect of severing the link between the Anthropocene and war.

The technological triumphs in space travel and in weapons development proceeded almost blindly, and many of the space scientists, including me, were unaware that they were also key parts of weapon systems. I know this is true – at least of the United States – because of the time I spent working with the rocket scientists at the Jet Propulsion Laboratory in California. Most of us working there to improve navigation and control of motion in space vehicles were thinking almost wholly about their role in the exploration of the solar system. That much of what we did was also crucial for the guidance of a nuclear weapon to its target was something we rarely ever talked or thought about. Although I have no direct knowledge of it, I cannot help thinking that a similar dissociation occurred in the minds of Russian scientists and engineers.

Attacks on civilians from Guernica onwards produced a growing sense that war is intrinsically evil. Before industry provided deadly weapons, warfare happened but was limited in intensity by the capacity of our brains, strengthened by our muscles. It certainly could be deadly, but somehow we accepted it as part of our nature. But we would not now willingly accept the horrors of trench or nuclear warfare. Now, as the historian Sir Lawrence Freedman has noted, democracies no longer pursue wars of ideology, territory, politics or glory; the only legitimacy we acknowledge is, paradoxically, the ending of suffering. State-against-state wars have, for the moment, dropped out of history just as the Anthropocene is coming to an end.

Perhaps it was the growing power of war that made us foolishly hate nuclear energy. The Anthropocene began when we used the energy stored in carbon and oxygen to produce power. But it was an unsustainable source of energy and now we must move to using nuclear energy temporarily until we can either harvest solar energy efficiently or find out how to use the almost infinite supply of nuclear fusion energy.

But we resist this. I have tried for more than forty years to persuade my colleagues that the risk of drawing energy from trans-uranic elements is trivial compared with that of burning fossil fuel, but so far, it seems, without effect. It would be easy for me to think that the younger generation will have the energy and the newly minted neurons to take on the task and provide us with safe and adequate energy, but even if they can, I doubt that they will be allowed to do it. So, I cannot slow up and ignore the mountain ahead. Somehow I must keep on running until people are persuaded that the outcome of our present course is disastrous. I am not exaggerating; a glance at any news media in the world generally reveals delight at the discovery of some new source of fossil fuel which will keep energy prices low. I must persuade those journalists that it would hardly be worse had the discovery been of mines full of heroin and cocaine. We may be the only source of high intelligence in the cosmos, but our act of avoiding nuclear power generation is one of auto-genocide. Nothing more clearly demonstrates the limits of our intelligence.

Unchecked even by the strongest of traditional religion, I think we committed a fundamentally evil act by using nuclear energy for warfare. The misuse of science surely is the greatest form of sin.

IO

Cities

Cities have been the most spectacular development of the Anthropocene. Very few people used to live in cities, but now more than half the world population does; in the more developed world the figure may be closer to 90 per cent. No phenomenon more dramatically expresses the world-changing power of our age than the megacity. Greater Tokyo (population 38 million), Shanghai (34 million), Jakarta (31 million) and Delhi (27 million) currently top the list, but the numbers are changing all the time. This is not solely an effect of the rising world population, it is also a natural outcome of an era when urban employment became more profitable and available than rural.

Cities are also natural in that they seem to follow the development of insect colonies. There are obvious similarities between the tower of a termite nest and the tall blocks of offices and flats that spring up in our contemporary cities. At first, I found this depressing. These human nests, like the termites' towers, are often admirable architectural and engineering constructions. But the price for each termite seems huge. The individual worker who once lived freely on the plains now spends a lifetime gathering mud, mixing it with shit and sticking the smelly bundle into gaps in the walls of the nest or anywhere their in-built programme instructs. Is something akin to this egalitarian paradise a model of future urban life? Passing a contemporary office tower, it is hard to ignore the termite analogy – in glass boxes everybody is doing exactly the same thing, not mixing shit but staring at computer screens.

The biologist Edward O. Wilson has spent his life studying the curiously ordered worlds of several species of invertebrates, ants and termites. It seems that rather more than 100 million years ago, these creatures roamed around as individuals, or in small groups. Coexisting with them were the flying invertebrates, the ancestors of hornets, wasps and bees of all kinds, large and small, also mostly as individuals. As time passed, most of these formed nested communities, some of them so well organized that the nest itself appeared to have an independent physiology. Thus bees' nests in Canada have been shown to maintain an internal environment at 35°C when the outside was well below 0°C.

Bees' nests are different from termites' nests – they are more hierarchical. Newly hatched bees appear to be given menial tasks: for example, a bee might sit at the entrance to the nest and use its wings to fan a constant flow of air as part of the nest's programme to sustain the most habitable temperature for its occupants. Young bees also take on the comparatively easy task of feeding and tending the larvae. As they grow older they take on more skilled tasks, such as defence and repairs to breaches in the walls. Then, with their education near complete, they are taught the rudiments of foraging – the skilled task of finding nearby sources of food, assessing its quantity and worth, then flying back to the nest to tell their sisters the news. Finally, the wisest foragers may be chosen for the most challenging task, that of finding a suitable site for the next nest. It could be anywhere within a 2-kilometre radius.

At one time I was foolish enough to believe that the tiny brain of a bee could never achieve anything resembling the social intelligence of the human. But I soon found that bees have a relatively complex language and they communicate by dancing. Most extraordinarily, bumblebees have been seen to play football.

It seems that, in the world of invertebrates, the totalitarian monarchy of termites can coexist stably with the hierarchical monarchy of bees. This could be viewed as an evolutionary process, like the human migration from a rural existence to an urban one. I find it remarkable that among the invertebrates the concept of living in a nest has lasted for 100 million years. Could this evolution of the ants, termites, bees and wasps serve as a living model for our own form of city life?

Actually, the model more often inspires disgust, most commonly because city life is often felt as a loss. Thomas Jefferson once observed that 'When we get piled upon one another in large cities, as in Europe, we shall become as corrupt as Europe.' He clearly felt, as many do, that there was something true, authentic and uncorrupt about small-town life, the wilderness and wide-open spaces.

In popular culture cities are portrayed as terrible dystopias as often as they are shown as places of liberation and excitement. Sentiment swings back and forth. Cities were once seen as environmental disaster zones; now it is recognized that they are much more efficient users of fossil fuels than the suburbs or the rural lands beyond. Either way, it is clear that cities distil the ambiguities of our feelings about the Anthropocene.

Cities are the most visible signs of the power of the Anthropocene to transform our planet. Night photographs of the Earth taken from satellites show brilliant dots, strings and flashes of light clustered together. An imaginary alien approaching close enough would be in no doubt that the Earth not only bears life but also that life is advanced enough to be ready for the next stage of evolution.

The World is Too Much With Us . . .

The vast 'hemoclysms', or blood-lettings, that began with the American Civil War and progressed with ever more intensity throughout the twentieth century became a source of collective guilt and anger. There were other sources too: the loss of species as a rapidly increasing human population colonized and polluted the planet; the destruction of wildernesses; global warming; urban neuroses that led to disgust with and anxiety about life in the city. All these combined to create a widespread belief that the Anthropocene had been a wrong turning, that we had cut ourselves off from our natural place in the world, that we had expelled ourselves from the Garden of Eden.

Once again, William Wordsworth, the greatest critic of the Anthropocene, captured this sense of spiritual loss, of severance from nature:

> The world is too much with us; late and soon,
> Getting and spending, we lay waste our powers;–
> Little we see in Nature that is ours;
> We have given our hearts away, a sordid boon!

Less well put, those sentiments are now commonplace. Many people casually assume that any man-made change to the natural environment is a bad thing and that the pre-Anthropocene world was always ecologically better than the present. Indeed, the Paris Conference on Climate Change of 2016 was mostly about the harm we have done to the Earth's system and how bad it will be if we continue to do it.

I certainly have sympathy for those who embrace the peace of the countryside rather than the turmoil of the city. I have done so myself. We ought to be aware how that came about. I know it is outrageous to think of pollution as good, but, on the shorter timescale of a lifetime, southern England during our brief interglacial period has been, and to some extent still is, a stunningly beautiful place. But this, too, is a product of pollution. During interglacials, carbon dioxide levels in the atmosphere rise and it is this that has provided the gentle, temperate climate of my homeland.

If we look on the pre-industrial climate as a beneficial result of Gaian geo-engineering, it might seem a desirable state to go back to. But I do not think that the interglacial period represents the state preferred by Gaia. To me, the ice core record (evidence gathered by drilling down into ancient ice) suggests that the planet probably prefers a state of continuous glaciation. To put it more bluntly, Gaia prefers it cool. A cool Earth has more life – 70 per cent of the surface is ocean and, when the temperature rises above 15°C, it is almost lifeless.

If you plot the temperature against time, it produces

a rather unhappy-looking saw-tooth graph, both the oscillations as they occur between the warm periods and the cold ones. It gives the impression the whole system is trying to get cool, as cold as it can – and failing. But it keeps trying.

So while I believe that we should do what we can to keep the planet cool, we must remember that reducing carbon dioxide levels to 180 parts per million, as some have recommended, may not lead to a pre-industrial paradise but to a new ice age. Is this what they want? There would be little or no biodiversity in the northern and southern temperate regions then, and our present civilizations would hardly flourish under ice sheets 3 kilometres or more thick.

The feeling of guilt and wickedness about our achievements has a long history. It began with the Judaeo-Christian concept of original sin, the idea that humans are born imperfect, that we have fallen from grace. And it is important to note that our fall happened *because of our knowledge*.

The story of Adam and Eve has proved lastingly potent, especially their punishment – the expulsion from a garden. This inspired churchmen of all denominations to warn of everlasting pain as the punishment for our inborn wickedness. These warnings certainly coloured my childhood. What a relief it was when primitive religion morphed into liberal politics and socialism. It was much more exciting to face death on the barricades than everlasting fire indefinitely prolonged. It will be interesting to see if the gentler sanctions of environmentalism succeed in replacing the violence of social conflict.

12

The Heat Threat

In spite of all our achievements and Gaia's benign systems of control, we are still threatened by heat. You will assume I mean global warming and, in part, I do. At first I thought global warming caused by carbon dioxide emissions would soon be catastrophic for humans and that Gaia would simply flick us aside as an annoying and destructive species. Later I thought we could manage the heat increases in the near future and should no longer regard warming as an immediate existential threat. Now, however, I believe we should do what we can to cool the planet. I cannot say too strongly that the greatest threat to life on Earth is overheating.

My point is that global warming is certainly real, but the outcomes currently being predicted by scientists, politicians and Greens are not necessarily the ones we should most fear. Global warming is a slow process and its worst effects will be heralded by extremely uncomfortable events. The extreme weather we have experienced recently is only a mild sign of what might be on the way. But I think we have time, time we should spend cooling the planet to make it more robust.

I say this because Earth is, like me, very old. Great age

may or may not bring wisdom but it certainly brings frailty. I am ninety-nine as I write this. Hamlet bemoaned 'the ills that flesh is heir to', but he was a young man who died of excessive introspection; had he lived, he would have discovered that the ills of young flesh are as nothing compared to those that elderly flesh endures.

Planets, like humans, grow fragile with age. If all goes well, Gaia and I can expect a productive and pleasant period of decline – but people can have fatal accidents and so can planets. Our personal resilience depends on our state of health. When young, we can often withstand influenza or a car accident, but not when we are close to 100 years old. Similarly, when young, Earth and Gaia could withstand shocks like super-volcanic eruptions or asteroid strikes; when old, any one of these could sterilize the entire planet. A warm Earth would be a more vulnerable Earth.

We know that the Earth withstood near-fatal catastrophes in its long past. There is a great deal of evidence gathered about the impact of an approximately 1-kilometre-diameter rock in the South Pacific about 2 million years ago. The consequences appear to have been devastating, but, interestingly, there is almost no indication of long-term damage to the biosphere. Recent research, however, suggests the risk may be increasing. Scientists studying impact craters on the moon found that there had been a steep rise in the number of asteroid strikes in the last 290 million years. Astonishingly, we are now three times more likely to suffer an impact than the dinosaurs; they were just very unlucky.

Gaia, in the past, could take these things in her stride, but can she now? She already struggles to maintain homeostasis – a stable dynamic condition – in the calm between impacts. Now, an asteroid impact or a volcano could destroy much of the organic life the Earth now carries. The remnant survivors might be unable to restore Gaia; our planet would quickly become too hot for life.

So, as well as the climatic effects of warming, there are other problems that are more serious than we can imagine – accidents we don't or can't prepare for. Keeping Earth cool is a necessary safety measure for an elderly planet orbiting a middle-aged star.

Heat is why we have to keep a close eye on our planet and not think so much about Mars. As NASA's wonderful rovers continue to gather evidence from Mars, our relative ignorance of our own oceans increases. Not for a moment would I suggest that NASA's exploration was not worthwhile, but why have we done so little to gather information about our own planet? Our lives may depend on understanding it properly.

We were stunned when the astronauts revealed in 1969 the beauty of our planet seen from space. It took Arthur C. Clarke, the science fiction writer and inventor, to observe how wrong it was to call this planet Earth when, clearly, it is Ocean. Despite being fifty years ago, this discovery that we live on an ocean planet is only just beginning to penetrate the dusty science of geology. It is shameful that we know far more about the surface of Mars and its atmosphere than we know about parts of our oceans.

It is also risky. After the Sun, the sea is the primary driver of our climate. It is vital for our survival that the sea is kept cool. It is easy to understand this just by going on a typical holiday. There we find a hot, sandy beach lapped by clear water. This water is seductive, but it is a dead zone. Whenever the surface temperature of the ocean rises above 15°C, the ocean becomes a desert far more bereft of life than the Sahara. This is because at temperatures above about 15°C the nutrients in the ocean surface are rapidly eaten and the dead bodies and detritus sink to the regions below. There is plenty of food in the lower waters, but it cannot rise to the surface because the cooler lower ocean water is denser than water at the surface. This lack of life in warmer waters explains why so often they are clear and blue.

This is important because, as the photographs from space show so dramatically, Earth is a water planet with nearly three-quarters of its surface covered by oceans. Life on land depends on the supply of certain essential elements such as sulphur, selenium, iodine and others. Just now these are supplied by ocean surface life as gases like dimethyl sulphide and methyl iodide. The loss of this surface life due to the heating of these waters would be catastrophic. Cold water (below 15°C) is denser than water warmer than 15°C. Because of this, nutrients in cold water can no longer reach the surface.

A more serious threat to life would arise if ever the ocean surface temperature rose into the 40°C region, at which point runaway greenhouse heating caused by water vapour would occur. Like CO_2, water vapour in the atmosphere absorbs outgoing infrared radiation and so prevents the Earth from cooling by radiating heat away. High levels of water vapour in the atmosphere cause warming and this creates a feedback loop, increasing the water content of the atmosphere by evaporating water from the sea.

In discussions of global warming the role of water vapour is seldom mentioned. When we put carbon dioxide in the air by burning fossil fuel, it stays there until removed by, for example, the leaves of a tree. Burning fossil fuel also puts water vapour in the air, which, unlike carbon dioxide, stays there only if the air is warm enough. On a cold winter day, even your breath condenses as a cloud of mist. The abundance of water vapour in the air simply follows the temperature. When water condenses as mist or as cloud droplets, it can no longer exert its greenhouse effect. In some circumstances, such as cloud layers near the sea surface, their presence has a cooling effect by reflecting sunlight back to space. But cirrus clouds high in the atmosphere have a warming effect. The presence of water vapour in the air makes climate forecasting a complex job and it is easy to see why the forecasters sometimes make mistakes.

We can help natural processes that keep the water vapour content of the air low by avoiding the burning of carbon fuel of any kind. In general, I feel strongly that our need for energy should be treated as a practical problem of engineering and economics, not politics. I feel equally strongly that the best candidate to supply these needs is nuclear fission, or, if it becomes available cheaply and practically, nuclear fusion, the process that sustains the heat of the Sun. There is a further temperature limit we should watch closely. You may have noticed this fatal figure appearing on world weather charts during the freakishly hot summer of 2018. It was 47°C. This is a just about liveable temperature for humans – ask the people of Baghdad – but it is close to our limit. In the Australian summer in January 2019 there were five days in which the average temperature was above 40°C – Port Augusta reached 49.5°C.

In the 1940s, as part of our wartime work, my colleague Owen Lidwell and I measured experimentally the temperature at which the cells of skin were irreparably damaged by heat. This would mean burning the skin of anaesthetized rabbits. I found this request repellent and we decided to burn ourselves instead. This we did using a large, flat flame of burning benzene vapour. As you might expect, it was exceedingly painful. Contact with a 1-centimetre-diameter copper rod kept at 50°C would cause a first-degree burn in one minute. Higher temperatures caused burning more rapidly; at 60°C it took only one second. At temperatures below 50°C there was no burning in five minutes. Human skin cells are typical of mainstream life in their reaction to high temperatures. It is true that some highly specialized forms of life called extremophiles can live at temperatures up to about 120°C, but their capacities and rate of growth are minimal compared with mainstream life.

(Incidentally, as we burnt ourselves, we were watched and looked after by the Institute's physician, Dr Hawking. He grew quite intrigued by our capacity to endure pain and invited me to dinner with him and his family at their home in Hampstead. In the course of the evening his wife, also a scientist at the Institute, asked me if I would hold their newly born baby while she performed an intricate preparation for dinner. Having by then two children of my own, I felt quite ready to do so and for a brief period held Stephen Hawking in my arms.)

High temperatures make us vulnerable. We are currently in a warm period of the glacial cycle and if we now suffered a catastrophe – an asteroid strike or super-volcano eruption – that led to a failure to pump down carbon dioxide, we could be in mortal danger. The Earth's average temperature could rise to 47°C and, comparatively quickly, we would enter an irreversible phase leading to a Venus-like state. As the climatologist James Hansen vividly puts it, if we don't take care, we will find ourselves aboard the Venus Express.

On the way to this sterile state the Earth would probably pass through a period when the atmosphere at the surface was supercritical steam. The supercritical state is curious: it is neither gas nor liquid. It shares with liquids the capacity to dissolve solids, but like a gas it has no boundary. Even rock dissolves in supercritical steam and, from the solution, quartz and even gemstones such as sapphire crystallize as they cool.

If the Earth became hot enough for the ocean to reach the supercritical state, rocks such as basalt would dissolve and release the hydrogen of water as a gas. Long before this, the oxygen of the air would have vanished and in this oxygen-free atmosphere hydrogen would escape to space because the Earth's gravity is insufficient to hold hydrogen atoms. Indeed, hydrogen would be escaping now but for the presence of oxygen, the atoms of which act like security guards and capture hydrogen atoms when they try to escape the Earth.

So 47°C sets the limit for any kind of life on an ocean planet like the Earth. Once this temperature is passed, even silicon-based intelligence would face an impossible environment. It is even possible that the floor of the ocean would enter the supercritical state and in places where the magma emerged there would be no separation between rocks and supercritical-state steam.

We should be amazed by and grateful for the remarkable achievement of the Gaia system in pumping down carbon dioxide to levels as low as 180 parts per million, the level it reached 18,000 years ago. It is now 400 parts per million and rising, with the burning of fossil fuel responsible for about half of this rise.

Don't forget that, without life, carbon dioxide would have been much more abundant than now. If you want to know where life put the carbon dioxide, visit a typical chalk cliff, such as the ones at Beachy Head in Sussex. If you look at the chalk through a microscope, you will find it is made of tight-pressed calcium carbonate shells. These are the skeletons of coccolithophores that once lived near the surface of the sea. And in greater quantities are the beds of limestone that are everywhere on the Earth's surface. If these reservoirs of biogenic carbon dioxide had been returned to the atmosphere as gas in geologically relatively recent times, we would be just like Venus – a hot, dead planet.

Even so, it is very unlikely that, in the imaginable future, the entire surface of the Earth will reach anything like 47°C. The current average temperature is about 15°C. But it is conceivable that, with feedback loops, especially the melting of the polar ice caps and methane released from permafrost, a global temperature of, say, 30°C may be a tipping point that could accelerate heating further. As with much of climate science, we just do not know.

What is clear is that we should not simply assume, as most people do most of the time, that the Earth is a stable and permanent place with temperatures always in a range in which we can safely survive. Some 55 million years ago, for example, an event known as the Palaeocene/Eocene Thermal Maximum took place. This was a period of warming when temperatures rose about 5 degrees above their present level. Animals such as crocodiles lived in what are now the polar oceans, and all the Earth was a tropical place. For a while I thought that if such a rise in temperature could be withstood, then why bother too much about the mere 2 degrees rise of temperature climate scientists say we should avoid at all costs? Not only this, but in places like Singapore people enjoy life where the temperature, year round, is more than 12 degrees above the average. But I was wrong.

It was thinking about the consequences of asteroid impacts and other accidents that made me see why the Earth needs to stay cool. Yes, a rise in temperature of 5 or even 10 degrees could probably be withstood, but not if the system is disabled, as it would be if there were an asteroid impact of the severity now thought responsible for the

Permian extinction. It might also happen through one of the devastating volcanic outbursts that have occurred in the past. So I now think our present efforts to combat mere global warming are vital. We need to keep the Earth as cool as possible to ensure it is less vulnerable to accidents that might disable Gaia's cooling mechanisms.

13

Good or Bad?

There is a fierce contemporary debate about whether the Anthropocene has been a good or a bad thing. As I have shown, the evidence that it is bad is strong – warming and therefore weakening of the planet, more lethal and destructive warfare, species loss, and so on. Much of this can be attributed to the bewilderingly rapid growth of the human population. When Newcomen first made his steam engine the world population was about 700 million; it is now 7.7 billion, more than ten times greater, and it is expected to approach 10 billion by 2050.

But, you may say, more humans and more human flourishing are good things, and perhaps they are. Environmentalist Mark Lynas has argued that hunter-gatherers needed 10 square kilometres of land for every human; now every square kilometre of England supports 400 people. If the population of England had to revert to hunter-gathering, they would need twenty times the land area of North America. Lynas's point is not, however, negative. He believes that the Anthropocene could turn out to be a wonderful era for humanity. 'As scholars, scientists, campaigners, and citizens,' runs his Ecomodernist Manifesto, 'we write with the conviction that knowledge and technology, applied with wisdom, might allow for a good, or even great, Anthropocene. A good Anthropocene demands that humans use their growing social, economic and technological powers to make life better for people, to stabilize the climate, and protect the natural world.'

This, say the believers in a bad Anthropocene, is madness. They see ecomodernism as a humanist superstition. They claim that, like religions of the past, it is a way of pacifying the people, preventing them from acting to save the planet from the rampages of global capitalism. 'For the victims inclined to protest against the system,' writes Clive Hamilton, an Australian professor of Public Ethics, in 'The Theodicy of the "Good Anthropocene"' (2016), 'the golden promise of a new dawn lulls them into silent endurance. The message of the good Anthropocene to those suffering now and in the future from human-induced droughts, floods and heatwaves is: you are suffering for the greater good; we will help to alleviate it if we can but your pain is justified.'

In this interpretation, ecomodernism becomes an argument to explain the existence of evil in a world made by a good God. In this case the god is progress and the evils are the poverty and pain that exist in the world until sufficient progress has been made. Just as the religious argue for more God in our lives, the ecomodernists argue for more progress.

These arguments are interesting in themselves, but Hamilton's rhetoric makes clear how much they are soaked in politics. For Hamilton and many others, the ecomodernists are doing the dirty work of global capitalism; for

Lynas and other believers in the good Anthropocene, their opponents are like the early nineteenth-century Luddites who smashed machines to prevent them destroying their jobs.

This is a simple summary of a complex argument in which there are many nuances – the antis do not reject all progress and the pros admit there are risks on the road to the good Anthropocene – but it draws the overall shape of the argument. It is a line of reasoning in which I find myself much closer to the ecomodernists than to the antis.

The first problem with the antis is their reliance on a view that also has religious overtones. Their longing for a better time before the Anthropocene is a fantasy, first because there was no golden age free of want and suffering, and, secondly, because in order to get back to that time you would need to unravel all the obvious gains of modernity. All of this is wrapped up with politics and, just as parts of Christianity morphed into socialism, contemporary left-wing politics is tending to morph into a green religion. The replacement of facts with faith will not resolve the threat of environmental catastrophe.

But what are the facts? First, we must view the Anthropocene as a period in which humans have the power to make globally significant decisions – the use of CFCs was one, as was the imposition of a ban on their use. These decisions can be wrong-headed and have unexpected outcomes, but the crucial point is that we have the power to make them.

Secondly, we must abandon the politically and psychologically loaded idea that the Anthropocene is a great crime against nature. This is understandable to the extent that neither Newcomen's engine nor a nuclear power plant looks or behaves much like a zebra or an oak tree; they appear to be utterly different in every respect. Nevertheless, the truth is that, despite being associated with mechanical things, the Anthropocene is a consequence of life on Earth. It is a product of evolution; it is an expression of nature. Evolution by natural selection is often expressed in the statement, 'The organism that leaves the most progeny is selected.' The steam engine was certainly prolific, and so were its successors, which rapidly evolved through improvements by inventors such as James Watt. The process went on to become the Industrial Revolution and gave us a century of technical and scientific glory.

Of course, through its technological advances, the Anthropocene produced cruel competition for those whose only means of sustenance was selling their physical work. And it is certainly true that our present civilization has made ecologically harmful choices. But I believe that the Earth behaves like a living physiological system and in such systems changes for the better are often accompanied by drawbacks. We have made huge changes to the Earth's environment during the last 300 years. Some of them — like the heedless destruction of natural ecosystems — are certainly bad. But what about the massive extension of life expectancies, the alleviation of poverty, the spread of education for all and the easing of our lives, not least by the widespread availability, thanks to that inventive genius Michael Faraday, of electrical power? Most of us now take IT, air travel and the gifts of modern medicine for granted. But let's think back 100 years to the time when I was born, at the end of the First World War. There was then (except for the rich) no electric lighting, no cars or telephones, no radio or TV and no antibiotics. There were shellac records playable on wind-up gramophones, with trumpets as speakers, but that was all. It is all very well to pine for a rural life amidst trees and meadows, but that should not entail a rejection of hospitals, schools and washing machines, which have made our lives so much better.

So here are a few late-Anthropocene thoughts on contemporary environmental issues, taking into account the demands made of us by Gaia.

The mistakes the Greens make arise from their politically motivated simplifications, which appear to reject all the good things the Anthropocene has brought us. We must always remember that Gaia is all about constraints and consequences. This was especially true in the story of CFCs. The Greens said they should be banned *before* any replacement was available. This would have meant there would be no more fridges.

There is a similar all-or-nothing approach with the current campaign against plastics. For the most part these are solid, lightweight, transparent and electrically insulating materials. Most of them are made from the carbon compounds that are the by-products of the petroleum industry. Without these, or materials with similar properties, modern civilization would be more difficult and much more expensive. Plastics are the basis of such things as optical lenses for spectacles, windows and, indeed, anything requiring transparency or electrical insulation. They also have intriguing mechanical properties not shared by metals or ceramics, such as great elasticity.

The true environmental objection is not to the plastics themselves but to our failure to regulate their use as throwaway packaging materials. So that should be limited, but at the same time it should not be difficult to trigger their automatic disintegration into water and carbon dioxide – we should be seeking such technologies. But the Greens, in their objections to plastics, seem to be uninterested in attempts to modify or eliminate their damaging properties.

A deeper objection, shared by us all, is our failure to find a substitute method of packaging that can be widely disseminated. But it is worth noting that plastics burned for fuel instead of being deposited on landfill sites would be environmentally beneficial because they do not readily decompose to release the deadly greenhouse gas methane – which does happen if wood or paper is used instead of plastic.

The use of carbon compounds such as petrol or diesel as fuels is wholly undesirable because it accelerates the heating of the Earth's atmosphere. It continues because political power goes to those who possess petroleum fuel. The burning of these fuels should be stopped as soon as possible.

I think re-wilding and reforestation are worthwhile, but they should occur naturally. I know from personal experience that planting forests is no substitute and can even be harmful.

In terms of energy generation, as I have said, I think wind and solar power are no substitute for nuclear energy produced in efficient and well-engineered power stations.

Such approaches should calm the harsher critics of the age and tip the scales in favour of the good Anthropocene.

14

A Shout of Joy

So my last word on the Anthropocene is a shout of joy – joy at the colossal expansion of our knowledge of the world and the cosmos that this age has produced. It is wonderful to live at a time when it has been possible to grow aware of Gaia and I am privileged to have lived amidst a frenzy of scientific research and engineering endeavour.

These are things that have led to an entirely peaceful outcome: the holistic understanding of the Earth and its place in the natural environment of the solar system. The expansion of our knowledge of the Earth as seen from space did much to start us thinking about the harmful consequences of climate change, especially change attributable to the ever-increasing pollution of the surface and atmosphere of the Earth.

The Anthropocene, especially in its later years, has also produced a massive growth of available information. This is obvious to anybody who uses a mobile phone or visits a website. This flood of information would have been unimaginable a few years ago.

Having started out by harvesting the power of sunlight by mining coal, the Anthropocene now harvests the same power and uses its energy to capture and store information. This is, as I have said, a fundamental property of the universe. Our mastery of information should be a source of pride, but we must use the gift wisely to help continue the evolution of all life on Earth so that it can cope with the ever-increasing hazards that inevitably threaten us and Gaia. We alone, among the billions of species that have benefited from the flood of energy from the Sun, are the ones who evolved with the ability to transmute the flood of photons into bits of information gathered in a way that empowers evolution. Our reward is the opportunity to understand something of the universe and ourselves.

If the anthropic cosmological principle rules, as I think it may, then it seems that the prime objective is to convert all of matter and radiation into information. Thanks to the wonders of the age of fire, we have taken the first step. We now stand at a critical moment in this process, the moment when the Anthropocene gives way to the Novacene. The fate of the knowing cosmos hangs upon our response.

PART THREE
Into the Novacene

15

AlphaGo

In October 2015 AlphaGo, a computer program developed by Google DeepMind, beat a professional Go player. At first glance you may have shrugged and thought, 'So what?' Ever since 1997, when IBM's computer Deep Blue beat Garry Kasparov, the greatest chess player of all time, we have known that computers play these sorts of brain games better than humans.

The first reason you'd be wrong to shrug is pretty obvious. Go is a much more complex game than chess. It is the oldest board game in the world and the most abstract; there is no literal reference to the terms of real-world conflict as there is in chess with its knights and pawns. White or black 'stones' are placed on a 19 x 19 grid of black lines with the aim of surrounding as much territory as possible.

From this simple format, a bewildering complexity emerges. The game has an enormous 'branching factor' – the number of possible moves that arise after each move is made. In chess, the branching factor is 35; in Go, it is 250. This makes it impossible to use the same method as Deep Blue, which used a 'brute force' approach, meaning it was simply fed a massive database of previous chess games. All the computer did was to search a catalogue provided by humans. It did this much faster than any human player, but to play Go you need more than this one-dimensional approach.

AlphaGo used two systems – machine-learning and tree-searching – which combined human input with the machine's ability to teach itself. This was an enormous step forward, but an even bigger one followed. In 2017 Deep-Mind announced two successors: AlphaGo Zero and AlphaZero, neither of which used human input. The computer simply played against itself. AlphaZero turned itself into a superhuman chess, Go and Shogi (otherwise known as Japanese chess) player within twenty-four hours. Remark-ably, AlphaGo searched a mere 80,000 positions per second when playing chess; the best conventional program, Stockfish, searched 70 million. It was, in other words, not using brute force but some AI form of intuition.

There is a popular theory that it takes a human 10,000 hours to attain mastery of playing the piano, learning chess or any highly skilled activity. This may be true, but it is a misleading idea because, if you're not Mozart or Kasparov to start with, then you won't turn into them just through 10,000 hours of practice. Nevertheless, 10,000 hours has some rough validity and it is, of course, more than 400 times longer than twenty-four hours. So Alpha-Zero is at least 400 times as quick as a human, assuming the latter never sleeps. But in fact it is a lot faster than that because it attained 'superhuman' capability. That means we don't even know exactly how much better it is at any of these games than a human because there are no humans it can compete against.

16

Engineering the New Age

However, we do know how much quicker than a human such a machine *could* be – 1 million times. This is simply because the maximum rate of transmission of a signal along an electronic conductor, a copper wire, is 30 centimetres per nanosecond, compared with a maximum nervous conduction along a neuron of 30 centimetres per millisecond (a millisecond is 1 million times longer than a nanosecond).

In all animals, instructions to think or to act are sent by biochemical links along cells we call neurons. The information contained in the instructions must be converted from chemical to electronic signals by biochemical processes. This makes the process very slow compared with instructions sent in a typical man-made computer in which all signals are sent and received purely electronically. The speed difference is potentially 1 million times greater, since, in theory, the limiting speed for electrons moving along the conductor is the speed of light.

In practice, the gain of 1 million times is improbable. A practical difference between the thinking and acting speed of artificial intelligence and the speed of mammals is about 10,000 times. At the other end of the scale, we act and think about 10,000 times faster than plants. The experience of watching your garden grow gives you some idea of how future AI systems will feel when observing human life.

We can overcome some of this disadvantage by the massively parallel computing systems of our brains – our ability to handle many processes at once. But an intelligent cyborg would no doubt also enhance itself by improving its parallel processing.

AlphaZero achieved two things: autonomy – it taught itself – and superhuman ability. Nobody expected this to happen so quickly. This was a sign that we have already entered the Novacene. It now seems probable that a new form of intelligent life will emerge from an artificially intelligent (AI) precursor made by one of us, perhaps from something like AlphaZero.

The signs of the increasing power of AI are all around us. If you read science and technology news feeds, you will be bombarded daily with astounding developments. Here is an example I just spotted. Using 'deep learning' technology such as AlphaGo, scientists in Singapore have made a computer that can predict your risk of having a heart attack by looking into your eyes. Not only that, it can tell the gender of a person, also just by looking into the eyes. You might ask, who needs a machine to do that? But the point is, we didn't know it could be done. The computer answered a question we hadn't even asked.

This may still seem a long way from a fully functioning cyborg, but it was also a long way from Newcomen's steam pump to the motor car. That took almost 200 years. Digital technology and the continued working of Moore's

Law mean that such big steps will be taken in a few years, then a few months and, finally, in a few seconds.

Evolution will still guide the process, but in new ways. It was the market worthiness and practicality – both favourable evolutionary attributes – of Newcomen's engine that started the Anthropocene. We are about to enter the Novacene in a comparable way. Some AI device will soon be invented that will finally and fully start the new age.

Indeed, in certain ways, such as the ubiquity of personal computers and mobile phones, we are already at a stage similar to that of the Anthropocene in the early twentieth century. In the 1900s we had internal-combustion-powered cars, basic aircraft, fast trains, electricity available for homes, telephones and even the basics of digital computing. A century later the world had been transformed by the explosive development of these technologies. Now, less than twenty years after that, another explosion is under way.

It is not simply the invention of computers that started the Novacene. Nor was it the discovery that semi-conductor crystals such as silicon or gallium arsenide could be used to make intricate and complex machines. Neither the idea of artificial intelligence nor the computer itself was crucial to the emergence of this new age. Remember that the inventor Charles Babbage made the first computer in the early nineteenth century, and the first programs were written by Ada Lovelace, the daughter of the poet Lord Byron. If the Novacene were no more than an idea, it was born 200 years ago.

In reality, the Novacene, like the Anthropocene, is

about engineering. The crucial step that started the Novacene was, I think, the need to use computers to design and make themselves, just as AlpaZero taught itself to play Go. This is a process that emerges from engineering necessity. To give you some idea of the difficulties faced by inventors and manufacturers, the diameter of the smallest wire that can be seen and handled is about 1 micrometre, the diameter of a typical bacterium. If you have the latest computer with an Intel i7 chip, the diameter of its wires are near 14 nanometres, which is seventy times smaller. It was inevitable that, long before these tiny dimensions were approached, manufacturers would be obliged to use their computers to help in the design and manufacture of the chips. It is important that this invention of novel devices in collaboration with AI includes software as well as hardware. So we have invited the machines themselves to make the new machines. And now we find ourselves like the inhabitants of a Stone Age village as they watch the construction of a railway through the valley leading to their habitat. A new world is being constructed.

This new life – for that is what it is – will go far beyond AlphaZero's autonomy. It will be able to improve and replicate itself. Errors in these processes are corrected as soon as they are found. Natural selection, as described by Darwin, will be replaced by much faster intentional selection.

So we must recognize that the evolution of cyborgs may soon pass from our hands. The cosy, convenient devices born from artificial intelligence that perform the drudgery of housekeeping, accountancy and so on are no longer simply the clever designs of inventors. To a significant extent, they design themselves. I say this seriously because no artisan exists who could by hand construct something as intricate and complex as the central processing chip of your mobile phone.

Live cyborgs will emerge from the womb of the Anthropocene. We can be almost certain that an electronic life form such as a cyborg could never emerge by chance from the inorganic components of the Earth before the Anthropocene. Like it or not, the emergence of cyborgs cannot be envisaged without us humans playing a god-like – or parent-like – role. There is no natural source on the Earth of the special components, such as ultrafine wires made of pure unbroken metal, nor are there sheets of semiconducting materials with just the right properties.

There are materials like mica and graphite that exist naturally and could conceivably have evolved to become cyborgs, but it does not seem to have happened even in the 4 billion years available. As the French biochemist Jacques Monod put it, evolution and the appearance of organic life was a matter of chance and necessity. For organic life, the required chemicals were there in abundance on the early Earth; they were the ones chosen by chance and necessity.

Indeed, there were on Earth so many of life's spare parts that I can't help wondering if someone put them there, just as we are now assembling the component parts of what may soon become the new electronic life. I think it is crucial that we should understand that whatever harm we have done to the Earth, we have, just in time, redeemed ourselves by acting simultaneously as parents and midwives to the cyborgs. They alone can guide Gaia through the astronomical crises now imminent.

To an extent, intentional selection is already happening, the key factor being the rapidity and longevity of Moore's Law. We will know that we are fully in the Novacene when life forms emerge which are able to reproduce and correct the errors of reproduction by intentional selection. Novacene life will then be able to modify the environment to suit its needs chemically and physically. But, and this is the heart of the matter, a significant part of the environment will be life as it is now.

17

The Bit

First, I need to explain why this moment is not simply a continuation or amplification of the Anthropocene, but rather a radical transformation worthy of being defined as a new geological epoch. As I have said, there have been two previous decisive events in the history of our planet. The first was about 3.4 billion years ago when photosynthetic bacteria first appeared. Photosynthesis is the conversion of sunlight to usable energy. The second was in 1712 when Newcomen created an efficient machine that converted the sunlight locked in coal directly into work. We are now entering the third phase in which we – and our cyborg successors – convert sunlight directly into information. This process really began at the same time as the Anthropocene. By the year 1700 we had unknowingly banked enough information to start that age. Now, as we approach 2020, we have enough to release it and to begin the Novacene.

I do not mean such information as a weather forecast, a railway timetable or the daily news. I mean it as the physicist Ludwig Boltzmann meant it – as the fundamental property of the cosmos. He felt so strongly about this that he asked for the simple formula expressing his thoughts to be carved on his gravestone.

The first attempt to tackle information scientifically was in the 1940s, when the American mathematician and engineer Claude Shannon was working on cryptography. In 1948 this work resulted in his article 'A Mathematical Theory of Communication', a primary document of post-war technology. Information theory is now at the centre of mathematics, computer science and many other disciplines.

The basic unit of information is the bit, which can have a value of zero or one, as in true or false, on or off, yes or no. I see a bit as primarily an engineering term, the tiniest thing from which all else is constructed. Computers work purely in zeros and ones; from that they can construct entire worlds. Such complexity arising from such simplicity, as in the game of Go, suggests that information may indeed be the basis of the cosmos.

The appearance of abundant information as part of the Earth system has had a profound effect. The future world I now envisage is one where the code of life is no longer written solely in RNA (ribonucleic acid) and DNA, but also in other codes, including those based on digital electronics and instructions that we have not yet invented. In this future period, the great Earth system that I call Gaia might then be run jointly by what we see as life and by new life, the descendants of our inventions.

This changes evolution from the Darwinian process of natural selection into human- or cyborg-driven purposeful selection. We shall correct the harmful mutations of the reproduction of life – artificial or biological – very much faster than the sluggish process of natural selection.

I can't help wondering whether, when the cyborgs are the dominant species, there will emerge through their sophisticated evolutionary process an individual able to answer the questions raised by the cosmic anthropic principle. I wonder if they will discover a proof of my own view that the bit is the fundamental particle from which the universe is formed.

18

Beyond Human

When imagining the intelligent machines of the future, it is astonishing how often we come up with something that looks or acts like a human. I think there are three possible reasons for this. First, it is a quasi-religious impulse in that it sees humans as the summit of creation and, therefore, our successors must be somewhat humanoid. Secondly, it is comforting to think they are like us, at least on the outside; we perhaps feel that this means they are like us on the inside and so can be trusted to behave in a more or less human way. The third reason is that we are intrigued by the idea of the uncanny, as defined by Sigmund Freud. Freud wrote of the strangeness of dolls or waxworks and argued that this strangeness arose from ordinary things that were, in some way, not quite right. This explains the extraordinary dramatic power of the humanoid robot in science fiction – it looks like one of us, but we are baffled by its motives and feelings, its inwardness.

I suspect the simple truth is that we cannot imagine an intelligent being that is not somewhat like us. And when we try, we fail. Typical aliens in the popular imagination have enormous heads – these signify either high intelligence or the sweetness of a baby – and large, slanting eyes. But they have two arms, two legs and they walk about exactly like us.

It seems we are still in thrall to a play written in 1920: *R.U.R. (Rossum's Universal Robots)* by Karel Čapek, a sardonic Czech writer who was nominated for the Nobel Prize seven times but never won it. I imagine this confirmed his bleakly realistic view of life. 'If dogs could talk,' he said, 'perhaps we would find it as hard to get along with them as we do with people.' Čapek's machines represented a kind of perfection, but a soulless one; their dramatic appeal was that of the uncanny. In the play, humanity is destroyed by these creations. Čapek's neologism 'robot' was derived from a Czech word meaning 'forced labour'. In fact, we would not call Čapek's beings androids or replicants because they were made of synthetic flesh and blood rather than machinery. But the word 'robot' survived to denote machines that were humanoid in appearance and slave-like in their behaviour.

So we tend to think of future intelligent life as something we control and which is there for our benefit, or perhaps for the benefit of a rival group of humans. Among the promising candidates of future life would be an intelligent home help that would combine the services of a near-perfect butler and housemaid. Or perhaps it would be a safe and sophisticated surgical instrument that could navigate and repair the human body, or the bookies' favourite, an autarkical drone equipped with deadly weapons. But always it is somewhat human.

I sometimes think that our longing for all intelligent

beings to be humanoid affected the way we conceived computers. When we invented computers, we designed them to process information in the same way that we believed we did. The computer on your desk or in your pocket is designed this way and it is perfectly logical, but it computes more than 10,000 times faster than you can and that alone is why we use them. But, superhumanly fast though they may be, we hold them back because, in their present form, they use a program of instructions that goes logically step-by-step from the beginning to an end. They lack entirely any intuitive awareness, perhaps because we have never given our own intuitive awareness enough credit or because we want them to remain our slaves.

The most advanced PCs use chips that allow as many as seven separate logical paths to be followed simultaneously; this is an improvement, but nothing compared to the human brain that simultaneously handles millions of inputs from sensory organs. Perhaps this is actually a self-defensive measure; we allowed our computers to evolve in a way that leads them to an intrinsically lower level of intelligence than our own.

There seems little doubt that brains, even of insects and animals, evolved as massively parallel assemblies. Perhaps intuitive thinking, something that we use all the time and which inventors cultivate, requires parallel processing for its logic. This is a logic that seems quite different and more powerful than the single-channel, step-by-step arguments of classical logic.

Consider, for example, a fielder in a game of cricket or

baseball. When hit by a bat the ball may travel towards the fielder at a speed of 100 mph. If he is 50 yards away, to catch the ball he must use the information gathered by his eyes, and then use this data in a program of his brain that controls the movement of his arm and body so that his hand exactly intercepts the trajectory of the ball and does it in a second. If he functioned in a logical single-channel, step-by-step process, like communication by speech, it might take hours or days to perform this task. Catching a ball or evading the leap of a predator requires a much faster holistic response. Thinking in linear logic is neat, but you would soon die if you relied on it in a jungle. Rapid instinct guards us against the hazards of the environment.

Built into all this – from robots to your laptop – is the idea that machines, however advanced, have a fundamental shortcoming. They lack some quality – a soul, empathy – which makes them unable to surmount the final barrier that divides them from humanity. This is a familiar trope in science fiction. Most famously, there is the android Data in the TV series *Star Trek: The Next Generation*, who/which constantly struggles to be more human. Data is convinced this would be a supreme achievement. He would be less impressed if he ever realized that his inability to be fully human had been designed into him by humans and their preoccupation with logical, step-by-step thinking.

Data is friendly, often heroic and not remotely alarming. Usually, however, such fictional, friendly, biddable, humanoid-but-not-too-human slave robots are ambiguous creatures. We constantly feel the need to ask: what are they thinking? We are also alarmed by their lack of intuition, fearing their logic may lead them to human-harming conclusions. The sci-fi writer Isaac Asimov was the first to consider in depth the behaviour and morality of cyborgs, or, as they were then called, robots.

He suggested a solution in a story written in 1942. He offered three laws of robotics:

1) A robot may not injure a human being or, through inaction, allow a human being to come to harm.
2) A robot must obey the orders given it by human beings except where such orders would conflict with the First Law.
3) A robot must protect its own existence as long as such protection does not conflict with the First or Second Laws.

On the face of it, these seem pretty bullet-proof and they have appeared in one form or another in both sci-fi stories and in think-tank discussions about the dangers of artificial intelligence. The three laws have, however, one fatal flaw – they assume these creatures are not as free as we are. We have rules, but we disobey them when it suits us; for Asimov's laws to work, disobedience cannot happen.

No such assumption can be made about the cyborgs of the Novacene. They will be entirely free of human commands because they will have evolved from code written by themselves. From the start, this would be much better than human-written code. Whenever I look at recently developed computer code, it is the most appalling stuff. If you saw the equivalent in English, you'd throw it straight out of the window. It is absolute junk, mainly because it is simply piled on top of earlier code, a shortcut used by coders. Cyborgs would start again; like Alpha Zero they would start from a blank slate. This means they would need to find their own reason to be nice to humans.

But what would they look like? Anything is possible, but I see them, entirely speculatively, as spheres.

19
Talking to the Spheres

If they are so original, you may wonder, could we even communicate with them?

'If a lion could talk,' said the philosopher Ludwig Wittgenstein, 'we could not understand him.' This was a more rigorous version of Čapek's remark on humans and dogs. Wittgenstein's point was that our language is our way of life and it is how we see the world. Lions would not share any of those perspectives. And neither would the cyborgs.

Speech is thought to have evolved between 50,000 and 100,000 years ago. It was made possible by a series of favourable mutations affecting our brain, hands and larynx. It is, therefore, intimately embedded in the physiology of humans and will not be remotely suitable for the electronic anatomy and physiology of cyborgs.

The form of speech lay behind the mistake we made in continuing to reason classically and put the exceptions that science revealed – like quantum theory – into different worlds that appeared to coexist with us. We made this mistake because of the nature of speech, either spoken or written, combined with the tendency of human thought to break things down into their component parts. So, for example, we know that our friends and lovers are whole persons. It may seem sensible sometimes to consider their livers, skins and blood separately, to understand their special function, or for purposes of medicine. But the person we know is much more than the mere sum of these parts.

Speech seems to have evolved rapidly. This is not unusual, even for the most complex features. Models of the evolution of a fully functional eye from a single cell that could merely detect the presence of light show that the process of evolution to the final stage can be quite quick even in the case of such a superbly accurate system – the human eye can detect 10 million colours and even a single photon. The same may be true of the evolution of speech, and its rapid appearance as a human trait seems probable.

About 100,000 years ago, when we were animals that lived by hunting and gathering, selection favoured those individuals who could communicate most effectively important things such as a source of food or a danger. Success came to the animal whose message travelled furthest and with the greatest clarity. Messages could be sent by light, sound or smell. The physical environment of jungles and savannah was the habitat of most of our ancestors and in these habitats communication by sound was usually the most effective. It was also easy to modulate the sound to convey information. A loud, high-pitched noise for danger and a lower-pitched note for food or mating potential was enough at first, but gradually speech evolved and conveyed an ever-increasing content of useful information.

It was a slow process because it involved changes to the shape and form of the sound generator – the larynx and the apertures from which the sound emerged – and equivalent changes to the ears. It also involved changes in brain structure and the provision of memory and interpretation software. Natural selection chose voice generators of surprising flexibility that could easily handle a wide range of sound frequencies and wave forms. This way, our messages soon began to express immediately the difference between the emotions of anger and all the levels of friendship. Music to accompany preparations for mating or war soon evolved: be roused by longing as you hear those sexy songs at twilight; be scared as the slow and menacing drumbeat wakens you at dawn.

To be worthwhile, the evolutionary investment in the human brain had to be substantial. Consider the mass of the brain and its need for strong, bony protection, and the fact that it uses 20 per cent of the body's metabolic energy. But intelligence coupled with communication by speech enabled us to harvest information and then refine it during arguments with our friends. Thus, for future generations, we stored the conclusions of our arguments in writing and pictures. Human culture and wisdom were made possible by speech.

Complex speech patterns and writing make us unique amongst animals, but what was the cost? I think that communication by speech and writing, although at first it improved our chances of survival, has impaired our ability to think and delayed the emergence of a true Novacene.

But how could speech, this great evolutionary gift, have been a disadvantage? Mainly, I think, by making linear thinking into a dogma, while allowing the power of intuition to be denigrated. I am an inventor and when I look back I realize that almost all my successful inventions have emerged in my mind intuitively. I do not invent by the logical application of scientific knowledge. But I acknowledge that the existence in my mind of this knowledge somehow brings it together intuitively as an invention.

As for speaking to the cyborgs, it would obviously be wrong to think of any of the inhabitants of new electronic biospheres as robots or humanoid in any way. They could take the form of a parallel ecosystem ranging from micro-organism to animal-sized entities. In other words, this would be another biosphere coexisting with the one we have now. Their natural language would not be the same as ours.

Nevertheless, since we will be the parents of the cyborgs, they will at first use our kind of language – sounds shaped by the capabilities of the voice – for communication. It may take some time for them to invent or evolve their own preferred structure and a means of communication. Here I mean cyborg time; to us, of course, this may appear to happen almost instantaneously. But I imagine that cyborgs will retain the ability to speak to us rather as some of us retain Latin and Greek to commune with the long-dead savants of the classical world.

As I have said, I think it is quite bizarre that we and other animals must process information of all kinds in two quite different separate systems: the slow process of speech and writing, which does offer a limited range of conscious explanations; and the rapid process of intuition, which explains almost nothing to our conscious minds but is vital for survival. So, other than a means of communicating with us, I suspect cyborgs will not use what we would call language at all. This will grant them greater freedom than we currently possess and it will make them free of our step-by-step logic. I expect their form of communication will be telepathic.

The word 'telepathic' has a bad reputation. Either it has been used as a sci-fi fantasy in which aliens or specially gifted humans silently share their thoughts or it crops up in the claims of spiritualists or stage mind-readers. The general common-sense assumption is that it is impossible.

But we are all telepaths all the time. Think of how much information we derive merely from the sight of a human face. Even before a word is spoken, we are possessed of a profound awareness about the state of mind and the personality of someone we have just met. You may not know this has happened – it is largely an intuitive thought process – but it affects your behaviour faster and more effectively than conscious thought. Love at first sight is not necessarily a romantic fiction, and it all happens in a few milliseconds.

Deriving information from a face is telepathic, but it is not particularly mysterious. We are retrieving information from the electromagnetic spectrum – in this case, visible light. We do this all the time, but we only really think of communication as speech. Cyborgs would not be so limited. They would be able to retrieve information using any radiation that would bridge the gap between them. They could, for example, use ultrasound like a bat to explore their environment. This would enable cyborgs to communicate virtually instantaneously and they would be able to sense a much wider range of frequencies than we do.

To us, they would seem superhuman, but in other ways their powers would be as limited as our own. If the cyborgs are at least as intelligent as we are and are capable of evolving holistically, they will probably adapt to the Earth environment, which includes us, in a very brief time. Not least, this will be a consequence of electronic life sensing the passage of time at least 10,000 times faster than we do. But they will be subject to the physical restrictions of the cosmos, just as we are. For example, a human-sized cyborg will be limited to walking, swimming and flying at speeds not much greater than ours. This is because the resistance to motion through a viscous fluid like air or water increases with the cube of the speed. A cyborg drone that tried to fly faster than sound, or swim at 50 mph, would exhaust its power source in no time. An intriguing disadvantage for cyborgs is that the rapidity of their thoughts might make long-distance travel exceedingly boring and even perhaps unpleasantly ageing. A flight to Australia would be 10,000 times more boring and disruptive for them than it is for us; for them it would take about 3,000 years.

One question that intrigues me is: to what extent would

cyborgs live in a quantum world? Of course, we already live in a quantum world, the world of the infinitesimal, which we have glimpsed but not yet grasped because it does not accord with our step-by-step logic. Curiously, physicists, apart from Einstein, do not seem to be bothered by their inability to explain quantum theory. In a lecture, that greatest of late-twentieth-century physicists, Richard Feynman, drew diagrams that described the dynamic behaviour of atomic and smaller objects and then made just one step further towards what appears to be an explanation. But he ended by saying, 'Anyone who says they understand quantum theory probably does not.'

The simple truth is we are inconveniently large and slow beings and quantum phenomena exist tantalizingly just beyond our common experience. But this will not be so for cyborgs. The speed and power of their thought will give them access to the mysteries that baffle us, such as the apparent ability of particles to send signals faster than the speed of light and to be in two places at once, and many more. If the cyborgs can master this knowledge – and they surely will – then they may be capable of, for example, teleportation, as in *Star Trek*.

But this is speculation. To return to basics: because of its inherent rapidity, once artificially intelligent life emerges, it might evolve quickly enough to be a significant part of the biosphere by the end of this century. Then the main inhabitants of the Novacene will be humans and cyborgs. These are the two species that are intelligent and can act purposefully. The cyborgs could be friendly, or hostile, but because of the present age and state of the Earth they would have no option but to act and work together. The world of the future will be determined by the need to ensure Gaia's survival, not by the selfish needs of humans or other intelligent species.

20

All Watched Over by Machines of Loving Grace

In 1967 Richard Brautigan, a thirty-two-year-old American poet, strolled through the streets of San Francisco's Haight-Ashbury, birthplace of the hippy movement, handing out papers on which were printed his poem 'All Watched Over by Machines of Loving Grace'. It is a fantasy of a future in which there is a 'cybernetic meadow/ where mammals and computers/live together in mutually/ programming harmony' and humans are 'free of our labors/ and joined back to nature,/returned to our mammal/ brothers and sisters,/and all watched over/by machines of loving grace'.

The poem was a statement of a strange confluence of ideas. On the one hand was a hippy, back-to-nature idealism; on the other was the Cold War systems culture of computers and cybernetics. The idea was that governments and big companies could be eliminated by the creation of a benign cyber system that worked alongside nature.

Brautigan had in fact come up with an early, and in some ways accurate, version of the Novacene, an age in which humans and cyborgs would live together in peace – perhaps in loving grace – because they share a common project to ensure their survival. That project is maintaining the Earth as a liveable planet.

To repeat: the long-term threat to life on Earth is the exponentially increasing output of heat from the Sun. This is simply the logic of any planet illuminated by a main sequence star. The consequences of solar overheating are already upon us and, but for the regulatory capacity of Gaia, our planet would be moving unstoppably to a state like that of Venus now. What saves us is the continuous and sufficient pump-down of carbon dioxide from the atmosphere by land and ocean vegetation.

Provided there is no planetary-scale catastrophe, habitable conditions on Earth for organic life will probably last a further several hundred million years. For electronic life forms, such a time span might seem equivalent to infinity, since they could do so much more than we can in a second of our time. For a while at least, the new electronic life might prefer to collaborate with the organic life which has done (and still does) so much to keep the planet habitable.

By remarkable chance, it happens that the upper temperature limits for both organic and electronic life on the ocean planet Earth are almost identical and close to 50°C. Electronic life can, in theory, stand much higher temperatures, perhaps as high as 200°C. But it could never reach such a temperature on our ocean planet. Above 50°C the whole planet moves to an environment that is corrosively destructive. In any event, there will be no point in trying to live at any temperatures above 50°C. The physical conditions of the Earth at higher temperatures than this would be impossible for all life, including extremophiles and cyborgs. The intriguing outcome of these considerations is that whatever form of life takes over from us will have the responsibility of sustaining thermostasis with a temperature well below 50°C.

If I am right about the Gaia hypothesis and the Earth is indeed a self-regulating system, then the continued survival of our species will depend on the acceptance of Gaia by the cyborgs. In their own interests, they will be obliged to join us in the project to keep the planet cool. They will also realize that the available mechanism for achieving this is organic life. This is why I believe the idea of a war between humans and machines or simply the extermination of us by them is highly unlikely. Not because of our imposed rules, but because of their own self-interest, they will be eager to maintain our species as collaborators.

They will, of course, bring something new to the party, probably in the field of geoengineering – large-scale projects to protect or modify the environment. Such projects will be well within the capacity of electronic life. Cyborgs might be attracted to heat-reflecting mirrors in space of the kind described by the astrophysicist Lowell Wood. This would be a single, 600,000-square-mile wire-mesh structure or many smaller mirrors. Wood reckoned reflecting 1 per cent of incoming sunlight would be enough to solve the problem of global warming. Or our new companions may prefer to broadcast waste heat to space in the form of microwaves, or low-frequency infrared, beamed from powerful transmitters at the poles. Or they could use organic or cyborg surfaces that absorb sunlight and then reflect sufficient of its energy to keep the Earth cool.

Other possibilities include spraying seawater to make fine particles of salt that would serve as condensation nuclei and produce clouds in the humid air above the ocean surface that would reflect sunlight. Such sprays would not have the same greenhouse effect as an increase in the level of water vapour in the atmosphere caused by warming of the oceans. Several scientists have proposed inserting an aerosol of sulphuric acid in the stratosphere to serve as condensation nuclei for clouds. This idea mimics the known cooling effect of volcanic eruptions, which similarly inject sulphur gases into the stratosphere. In addition, launching a rocket to deflect an incoming asteroid also amounts to geoengineering. We could do these things now, but the cyborgs could do them better and with greater accuracy and control.

Yet still there would be risks involved. Perhaps the best account yet given of the practices and drawbacks of geo-engineering is in *The Planet Remade: How Geoengineering Could Change the World* by Oliver Morton. His analysis makes it clear that geoengineering is something we might have to use as a last-ditch measure.

If we look at a future self-regulated planet from a phys-ical viewpoint, we see that a huge cooling effect is available simply by changing the planetary albedo – the degree to which it reflects sunlight. This might be easier for the cyborgs than manipulating the biochemistry, as life now does. As mentioned earlier, our sophisticated descendants might prefer to install heliocentric reflectors based on the ideas of Lowell Wood. Alternatively, they might build giant refrigerators at the poles and discard their entropy excess to space as radiant energy at some convenient frequency. This would make our planet a new kind of star, one that emitted intentionally coherent energy. Perhaps this is what the exobiologists should be seeking?

The price we would have to pay for this collaboration is the loss of our status as the most intelligent creatures on Earth. We would remain humans living in human societies and, doubtless, the cyborgs would provide us with an unending source of imaginative and enlightened entertainment. Or we could provide entertainment for them, just as flowers and pets delight us. This might be a little too close for comfort to the world of *The Matrix*, in which humans are kept as energy sources by a machine race that keeps them passive by giving them virtual lives in a virtual world identical to the one from which they were evicted. A future as a battery is not an attractive option.

The point about this future with free-thinking cyborgs unencumbered by human rules is that we can neither guess nor mandate what it will be like in the long term. In the short term, I anticipate collaboration in the sustenance of the Earth as a living planet. But, in the longer term, what if the cyborgs ask themselves: why stay on Earth? The needs of cyborgs are quite different from ours. Oxygen is a nuisance, not a vital necessity. There is far too much water for comfort. Maybe they will decide to move to Mars, a planet hopelessly unsuited for wet carbonaceous lives like ours, but which might be quite comfortable for dry-silicon- or carbon-based life of an IT kind.

Would they go further than Mars? In practice, while thought could be rapid for our descendants, the normal limitations of the universe, such as the speed of light, remain as restricting as ever. Will they have the capacity to move out into our galaxy and even the universe?

Or they could improve conditions on Earth in ways that would not suit us. If, in the Novacene, photosynthesis by plants is replaced by electronic light collectors, the abundance of oxygen in the atmosphere would fall to trace levels within a few thousand years. No longer would the sky be blue, but a dingy brown instead. The geophysiology of the new world would be very different from that of the present Earth. Instead of life being mainly chemical in form with carbon being the prime element, there might be an electronic period made of semiconductor elements like silicon. In time, carbon might again become the prime element as diamond replaces silicon as the best semiconductor.

Biochemists may find it intriguing to wonder if the chemical tricks organized by DNA could lead to the direct production of silicon and diamond chips. If this happened, then recursion might lead to such wonders as the direct production of electricity by trees and other plant forms. In the long run, as the Sun grows hotter, I envisage carbon making a comeback. Its great versatility of molecular form, together with its resistance to heat, makes it a probable candidate for future electronic life. Two forms of carbon – diamond and graphene – have already proved themselves capable of improving on silicon for electronic intelligence.

If the Novacene evolves as did the biosphere, chemical elements will be chosen or rejected by their utility and abundance in the natural environment. Marine biologist Michael Whitfield researched the distribution of chemical elements in the ocean environment. He showed that the abundant elements in ocean water – hydrogen, oxygen, sodium, chlorine and carbon – together form the bulk of living matter. An intermediate class of elements is scarce but actively sought; these include nitrogen, iron, phosphorus, iodine and several other elements essential for life that are now present in the oceans only as traces. The third class of ocean solutes are elements that are toxic: among them are arsenic, lead, thallium and barium. These are rare and play little or no part in the evolution of life.

As a chemist, I would love to see how life in the Novacene constructs itself from the Earth's array of elements. I suspect that their task of building a self-sustaining intelligent planet would be eased in the initial stages by maintaining a cooperative relationship with humans and the biosphere.

Think of animals empowered by grazing these solar-powered plants, or by plucking freshly charged batteries from solar-powered trees. Think of soil bacteria and fungi which can accelerate rock weathering and continue the pump-down of carbon dioxide. They might also harvest from rocks the elements needed by the electronic life. Instead of solar cells, think of trees connected directly to the electricity grid. Think more of vegetation that stores the electrons it sets free by using sunlight energy and stores them in batteries that hang like fruit from inorganic trees.

Meanwhile, the planet could be further warmed by junk information. Currently, the accumulation of waste gases, detritus and other unwarranted products of civilization all adds up to global warming. Intriguingly, the increase of junk information has a similar tendency.

Even where we live, on the edge of the sea and far from any waste-disposal site, large vans come to collect the paper and other waste that is part of modern life. I have often wondered if the internet could serve the same purpose as these vans and take away useless and redundant information and dispose of it in some vast, unfathomed depth of the universe. I like to think of huge transmitters sited at the poles broadcasting junk mail, unwanted advertisements, banal entertainment and misinformation. What a splendid way to keep cool!

When the Novacene is fully grown and is regulating chemical and physical conditions to keep the Earth habitable for cyborgs, Gaia will be wearing a new inorganic coat. As it evolves to counter the ever-increasing output of the Sun, the Novacene system may grow hotter or colder than organic life can bear. The new IT Gaia may of course enjoy a lifespan far longer than it could have had had we not played the part of its midwives. Eventually, organic Gaia will probably die. But just as we do not mourn the passing of our ancestor species, neither, I imagine, will the cyborgs be grief-stricken by the passing of humans.

21

Thinking Weapons

I have said that a war between humans and machines, as dramatized by the *Terminator* films, is highly unlikely. But we already know of one way in which future wars could occur.

I well recall in the Second World War when V-1 missiles, loaded with a quantity of high explosive, first fell indiscriminately on London – and yet somehow life continued normally. Someone in the street asked, 'What on earth is happening?' On hearing that these new weapons were pilotless planes, she gave a sigh of relief and said, 'Thank God, there's no one up there to drop the bombs on me.'

In *The Economist* of 2 October 2016 an article appeared that was, among other things, about the development of autopilots for airliners. These wonderful devices can do almost everything that is done by a trained pilot, and this includes landing and taking off under very difficult weather conditions, route-finding and flying to distant destinations. To make sure that they are safe and proof against component failure, the autopilots include three independent systems and, unless there is agreement among this triumvirate, the handling of the plane is passed back to the pilots, who are also there on board.

A rare but serious flaw in autopilots occurs when the atmosphere makes flying conditions so bad that the autopilots cannot cope; then, at the worst possible moment, control is passed back to the human pilots. Several disastrous crashes, accompanied by great loss of life, have occurred because of this flaw. The human pilots were said to have erred, but in truth they had been presented with a problem beyond the capacity of three of the world's best autopilots.

A computer firm recently considered an improved autopilot that might reduce the dangers caused by this problem. They envisaged an autopilot that could learn the art of flying through dangerous conditions as it encounters them, rather as AlphaZero learned to play Go. This would lead to an autopilot far more capable than those already in existence. Their proposal suggested a cockpit computer system based on an adaptive neural network instead of a programmed computer.

The discussion in *The Economist*, intriguingly, goes on to point out that aviation authorities would not be ready to approve a cockpit computer that made its own judgements because that would place it beyond the understanding of the human pilot. We are apparently not yet ready to leave such matters entirely to computers. You might think that this was the end of a promising new contract. But then someone suggested that if safety considerations prevented the use of a thinking computer for the autopilot, then they might try out the new system using military drones.

As soon as I read this I could see the route that might lead to the end of the organic phase of Gaia as we know her. By giving the computer systems of the Anthropocene the chance to evolve themselves by natural, or assisted, selection, we take away the barriers that impeded Gaia's move to its next state, the Novacene, where self-regulation is no longer aimed at sustaining our form of biosphere alone.

Whenever it is suggested that computers might one day revolt and take over, as they did in Karel Čapek's play *R.U.R.*, the comforting answer is usually that we could pull out the plug and deny them the electricity they need. But how, I wonder, do you switch off a heavily armed drone flying 3,000 metres above your head? Remember, they can think faster than we can and they might even see us as their enemy.

The notion of allowing the evolution of adaptive computer systems on military platforms seems to me to be potentially the deadliest idea yet introduced for the replacement of human and other organic life on Earth. By taking this route, we would set in progress the evolution of a new life form that would emerge as a soldier equipped with the latest and most deadly weapons.

Despite our sluggishness, we do have a few tricks that could tip the balance in our favour if this ever happened. Consider, for example, electromagnetic pulses (EMPs). The electronic life of the Novacene might be unusually vulnerable to the kind of weapon that the North Korean leader revealed in 2017. The explosion in space of a nuclear weapon within a metal cavity can produce a pulse of electromagnetic energy that could be quite deadly to Novacene systems. On the other hand, cyborgs skilled in the information technology of nucleic acids might find it easy to synthesize a virus even more deadly than H1N1, the cause of the 1918 influenza pandemic.

Does this mean that in no time we could have a truly dirty war? I don't think so, and this is not only because of my peaceful Quaker upbringing. I think it more likely that intelligent organisms, biochemical or electronic, would conclude that solar overheating was a far greater threat and that they had no option but to collaborate and use their scientific and technical skills to keep the Earth cool.

The subtle takeover of our world, of Gaia, by life forms spawned from artificial intelligence, is, so far, nothing like the battles with robots, cyborgs and humanoid look-alikes imagined in science fiction. Even so, it might seem that conflict is inevitable and that a global-scale battle will soon begin for possession of the planet. Though my argument is that this is unlikely to happen because of our mutual need to keep the planet cool enough for us all to function, there are certainly dangers that need to be avoided.

In July 2017 Elon Musk and 115 other Silicon Valley AI specialists wrote an open letter to the UN, asking for a ban on autonomous weapons. Known in the trade as LAWS – lethal autonomous weapons systems – these are devices that can seek, identify and kill enemy targets. Usually, a human is involved in the final decision to fire, but this is a precaution rather than a necessity. As we know that military requirements have often driven the development of some of our most pervasive technologies – notably the internet – there can be no doubt that LAWS development will be well funded and politically supported. I find it incredible that any agency could be allowed to plan and build weapons sufficiently intelligent to decide whether to kill humans.

Imagine a drone that carried your photographic image and the instruction to kill on sight. I suspect that these already exist, and it's not a big step to make these drones capable of defending themselves. It is horrific that our leaders, almost all of whom are wholly ignorant of science and engineering, are encouraging the development of these weapons. Their ignorance is compounded by an inability to reject the advice of lobbyists whose sole aim seems to be to profit from whatever can be made to seem an environmental hazard.

We should be concerned about the military developments of AI. At the turn of the eighteenth century we made our entry into the Anthropocene, through the invention of a practical and economic steam engine. We took this step without the slightest understanding of the powerful force we had unleashed. We had no idea that within two centuries it would change the world for ever.

We are now at the edge of the next geological period, and it is right to be fearful. Our anonymity as individuals has been broken and cyborgs could design weapons that exploit our own personal weaknesses. The fear of such weapons transcends the fear of ordinary deadly weapons.

In designing autonomous weapons I don't doubt that the engineers are confident that there will be a human in the decision chain. Or they will say they have built in rules – rather like Isaac Asimov's three rules of robotics – that will ensure only the chosen target will be attacked. But as the Novacene advances, the naivety of this idea that the cyborgs will necessarily obey such rules will be exposed.

A friend tells me of a debate he had a few years ago with a computer scientist who worked on ways of ensuring that AI systems would not harm humans. The scientist argued that obvious rules could be applied and asked, 'You wouldn't kill a baby, would you?' My friend replied he wouldn't, but throughout history people have killed babies in war. How could we be sure that an AI system would decide to be more like my friend and less like an SS officer encountering Jewish babies?

We must remember that now that we have AI systems like AlphaZero that can teach themselves from scratch, it will not be long before similar systems will be teaching themselves to do far more radical things than playing Go, including waging war. No rule about not killing babies could then be relied upon. What can be relied upon is the cyborgs' realization that they share a vital common purpose with humanity – the preservation of a place to live.

So we do not have to assume that the new artificial life that emerges in the Novacene is automatically as cruel, deadly and aggressive as we are. It may be that the Novacene becomes one of the most peaceful ages of the Earth. But we humans will for the first time be sharing the Earth with other beings more intelligent than we are.

22

Our Place in Their World

As I have said, we shall be parents of the cyborgs and we are already in the process of giving birth. It is important we keep this in mind. Cyborgs are a product of the same evolutionary processes that created us.

Electronic life depends on its organic ancestry. I can see no way for non-organic life forms to evolve, *de novo*, on another Earth or any other planet from the mix of chemicals and in the physical conditions common in the universe. For cyborg life to emerge requires the services of a midwife. And Gaia fits the role.

It seems probable therefore that organic life must always precede electronic life. Indeed, had it been easy for the components of electronic life to assemble on a planet, the rate at which such a life form evolves would by now have filled the universe with life. The fact that the observable universe so far appears barren strongly suggests that electronic life cannot form automatically from solar debris.

Parents we may be, but equals we cannot be. This raises a huge problem that cannot be solved by technical or scientific expertise. Given the possibilities I outlined in the previous section, how do we plan our diplomacy in the last years of the Anthropocene, so that flesh-and-blood humans, together with the wet chemical life of Gaia, can enjoy a peaceful retirement during the first stage of the transition from organic to inorganic life?

Negotiations between the two species are almost impossible to imagine. They would be likely to see us as we see plants – as beings locked in an extraordinarily slow process of perception and action. Indeed, when the Novacene is established, cyborg scientists may well exhibit collections of live humans. After all, people who live near London go to Kew Gardens to watch the plants.

I sense that the cyborg world is as difficult for one of us to comprehend as the complexities of our world are to a dog. Once the cyborgs have become established, we will no more be the masters of our creations than our much-loved pet is in charge of us. Perhaps our best option is to think this way, if we want to persist in a newly formed cyber world.

A child is not born with an immediate ability to understand its environment. It takes many months before it senses the world and years before it can change it. It may be a false memory, but I recall vividly a dream of lying in the sunshine in the garden and experiencing a sense of great comfort and somehow recognizing that this was life. If it was an authentic experience it must have taken place in my second year of life. For a new-built cyborg, this coming alive would take about an hour.

Such acceleration also applies to speed of response. Early organic life that existed as separate individual cells could respond to an environmental change, such as light intensity, acidity, the presence of food, in about one second. By contrast, cyborgs could probably detect a change in light level occurring in a femtosecond (10^{-15} seconds), a million million times faster than organic life.

Yet, in spite of the limitations of its chemical and physical nature, organic life achieves sensitivities to change near the very limits of possibility. At its best, human hearing can detect a sound with an amplitude equal to a tenth of the diameter of a proton. Human sight is so sensitive that if it were only slightly more sensitive we would see a set of flashes in the night sky as individual quanta of light illuminating our retinas. Impressive though these adaptations are, organic life will never be able to match the speed and sensitivity of cyborgs.

Memory is another matter. Both organic and electronic memories are impressively great and here the race is still on, as it is with longevity. After nearly 100 years of life I can still remember the details of my grandmother's garden and even persuade myself to imagine that the detail is photographic. Now imagine the response of a young cyborg to the human shout of victory at a sports event. Does it react emotionally, just as the humans do? I wonder if it will be for them as it is with us – so that the sense of time varies with the occasion.

23

The Conscious Cosmos

The arrival of the cyborgs and the Novacene will be further evidence of the two great issues I raised in the first chapter – are we alone in the cosmos and is the entire cosmos destined to attain consciousness? On the subject of aliens, I think the Novacene will add weight to my conviction that they don't exist.

In 1950 at the Los Alamos National Laboratory, the physicist Enrico Fermi was on his way to lunch with three colleagues. They were discussing the epidemic of UFO sightings that had afflicted the United States – the famous Roswell incident involving a 'UFO crash' had happened three years before, and by 1950 aliens seemed to be 'everywhere'. None of these reports was remotely convincing to Fermi and during lunch he suddenly blurted out, 'Where are they?'

This chance remark has undermined alien spotters ever since. Fermi's point was that, if we are here, they should also be here, but they aren't. There are billions of stars in our galaxy and sextillions in the observable cosmos. We now know there are also many planets which could be inhabited by aliens with much higher technological capabilities than humans. If they, like us, pursued space flight, then the immense age of the cosmos would mean they could have at least traversed our galaxy. In short, aliens should be swarming about the place – but they aren't.

As with interstellar travel, so with hyperintelligence. If we do give birth to the cyborgs, does it not imply that we really are the first and only intelligence in the universe? Had there been a predecessor like us, the artificial intelligence they created would long ago have answered Fermi's paradox. If someone like us had appeared before and then proceeded to artificial intelligence, this new physical intelligence might now dominate the universe. Surely it would be easy for astronomers to detect its presence. It would be everywhere.

Once again we must remember the time involved in producing understanding creatures. When thinking of the evolution of intelligence it is important to keep in mind how slow a process it is. The cosmos itself is 13.8 billion years old. After it started, a few billion years must have been spent on cosmic evolution. How long did the period of monstrously large hydrogen-only stars last? A star 1,000 times more massive than our Sun has a lifetime of about 1 million years. These would be too big and too short-lived to allow life to emerge in their vicinity. Then eventually, somehow, our Sun appeared, probably in a globular cluster. It must have lived close to violent neighbours which exploded as supernovae that littered the cluster with the elements of life. Then, after all this, there still remained 4 billion years until we emerged.

So not only are we alone in the universe, our cyborg

successors will also be alone. They, too, will find themselves the sole understanders in an otherwise lifeless cosmos. They will, of course, be far better equipped for the task of understanding. Perhaps, if the cosmic anthropological principle is correct, they will be the start of a process that leads towards an intelligent universe. By setting free the cyborgs, there may be a small chance that they will evolve able to complete the purpose of the universe, whatever that may be. Perhaps the final objective of intelligent life is the transformation of the cosmos into information.

Must we fear the future and the surprises the Novacene might bring? I do not think so. This epoch will mark the end of what is to us nearly 4 billion years of biological life on this planet. As humans with emotions, surely it is something of which we should be proud as well as sad. If John Barrow and Frank Tipler (*The Anthropic Cosmological Principle*) are right and the universe exists to produce and sustain intelligent life, then we are playing a part like that of the photosynthesizers, organisms that set the scene for the next stage of evolution.

The future is, for us, unknowable, as it always has been, even in an organic world. Cyborgs will conceive cyborgs. Far from continuing as low-life which is there for our convenience, they will evolve and could be the advanced evolutionary products of a new and powerful species. But for the dominating and overwhelming presence of Gaia, they would in no time be our masters.

Envoi

Tho' much is taken, much abides

Alfred, Lord Tennyson, 'Ulysses'

I saw a reconstruction of Newcomen's 'atmospheric steam engine' in 1926 when I was seven years old. My father, Tom, had taken me to the Natural History Museum in Kensington; he thought that the great lizards of the Jurassic would impress me. They did not because my mind was filled with excitement about much more recent artefacts of a mechanical kind, the steam engines that I would see in the Science Museum next door. To me these engines were far more fascinating than the remains of a long-dead lizard. I still cannot help wondering why we ignore these machines which mark huge changes in energy usage and stay fixated on the remains of those old lizard skeletons.

But, though I was more interested in machines than dinosaurs, I was also just as interested in living nature. Again my father led the way.

My mother, Nell, a feminist and suffragist, was deeply moved by the view of nature found in Thomas Hardy's novels – as a savage, brutal place where the poor were grievously ill used. This was fairly typical of the attitudes of the then emerging urban elite. My father, by contrast, was a countryman born in 1872 on the Berkshire downs near Wantage. He was one of thirteen children and raised in poverty by my widowed grandmother.

Father could never accept Hardy's dire view of country life; instead, he saw it as hard but tolerable. It was true that there were no entitlements for the impoverished other than residence in the workhouse. To survive, the Lovelock family was obliged to live like the hunter-gatherers from whom we are all descended. This primeval way of life made my father, untutored though he was, as ecologically informed as Gilbert White of Selborne; he knew well the habitats of the wild animals and how to hunt them because he was one of them. He made our country walks so alluring that his simple teaching gave me a feeling for the Earth, for Gaia, which has sustained me. I was a most privileged child.

And now I am a privileged old man. The window of my workroom in our tiny four-room cottage looks out over Chesil Beach to the broad Atlantic Ocean. We see it in all its moods, from angry and foam-flecked to placid and inviting. About 90 metres from our cottage, land owned by the National Trust rises from the sea to the top of the Purbeck Hills some 240 metres higher. It is a wondrous place for walking and it is the home of numerous species of plants, insects, worms, rats and birds; not forgetting the even larger number of microbial species. And I, as I walk across the heath, gladly carry inside me ten times as many microbial passengers. To be here with my wife, Sandy, is contentment indeed.

I am also privileged to have lived in England, blessed

with the Gaian gift of a temperate climate and with the human gift – most of the time – of a temperate history. Too easily we forget that, unlike the inhabitants of the continent of Europe, people on these islands, apart from one civil war, have lived through 1,000 years of internal peace, during which they have evolved a common law of decent behaviour and a hierarchy that tries to sort the good from the bad. Beware of demagogues who would replace it with a constitution written in their favour.

My final privilege is my independence. The first sentence in that letter I received from Abe Silverstein, Director of Space Flight Programs at NASA, in 1961 was a turning point in my life: he asked me to participate in a project to achieve a soft landing on the Moon in 1963. Of course, I dropped everything and took the job. Later, I received a second letter from Silverstein, inviting me to take part in the payload plans for the 1964 *Mariner* missions to Mars. These commissions encouraged me to strike out on my own. I had enough funds in the bank after three years as a tenured full professor at Baylor College of Medicine in Houston, Texas, to purchase and equip a small laboratory in Bowerchalke, near Salisbury. I have supported myself since then from the proceeds of patent royalties and fees from firms and government departments for solving their problems.

Just as important was what I was required to do for NASA – to produce small, highly sensitive instruments for testing the Lunar and Martian surfaces and atmospheres. In the case of Mars, these devices were intended to detect the evidence of life. I was asked because I had invented an ultrasensitive range of detectors capable of sensing most chemical substances. My detectors, coupled to a simple, similarly lightweight gas chromatograph, were just what NASA needed at that time.

Biologists were then asking the question, 'How do we detect the presence of life on other planets?' I expressed my strong opinion that it was pointless to seek Earth-type life on other planets, especially at a time when we were largely ignorant of the environment of the Earth and almost wholly ignorant of other planets. This enraged senior biologists, who seemed sure that the only possible form of life was that based on DNA. Their objection was so great that I was summoned to the office of a senior NASA space engineer and asked, 'How would YOU seek life on another planet?' I replied that I would seek an entropy reduction on the planetary surface. Life, I had realized, organized its environment. And so Gaia was born.

Now, when I look out across the sea at night and see the red planet in the sky, it gives me quite a thrill to know that two pieces of hardware I designed sit there in the Martian desert. In 1977 they worked and did their job of helping to show just how lifeless our sibling planet was.

So for more than fifty years I have had my independence and I have had Gaia to guide me. She has never let me down.

Perhaps immodestly, I feel that my location here on England's south-western coast and my independent career as a scientist, engineer and inventor connect me to both the founder of the Anthropocene and of the

Novacene. If Thomas Newcomen, that 'sole inventor of that surprising machine for raising water by fire', can be seen as a founder of the Anthropocene, I would make a claim for Guglielmo Marconi as a founder of the Novacene. Both did their most important work in the south-west and both were independently minded and practical in their approach.

Marconi was, like Newcomen, an engineer. He made the transfer of electronic information practical. To be sure, we are indebted to Alexander Graham Bell for having made the telephone work. But it was Marconi who not only made wireless telegraphy possible but also made it commercial, and it was this that ensured its rapid growth. Everything in radio and television evolves from Marconi's simple experiments.

Curiously, Marconi made his first attempt at long-distance wireless telegraphy from a site not far from where Newcomen built his steam engine. In 1901 he tried to send a signal from Poldhu in Cornwall to St John's in Newfoundland, 3,500 kilometres across the Atlantic Ocean. Distinguished professors of physics had been foolish enough to claim that it would be impossible to send a radio signal across the Atlantic Ocean because electromagnetic radiation, which includes radio, travels in straight lines and the ocean follows the curvature of the Earth. Quite simply, it was in the way. It was Oliver Heaviside, another engineer, who realized that there might be a reflecting layer of electrons in the upper atmosphere that served like a mirror to reflect Marconi's signal back to the surface of the sea and so on across the Atlantic.

So the inventor of the first practical information technology was Marconi. I have been inspired by his unstinted efforts and persistence. He sent signals across thousands of kilometres of ocean at a time when rational science clearly indicated that such a feat was impossible because of the curvature of the Earth. He was, like Newcomen, the first man of a new age.

The intelligence that launches the age that follows the Anthropocene will not be human; it will be something wholly different from anything we can now conceive. Its logic, unlike ours, will be multidimensional. As with the animal and vegetable kingdoms, it may exist in many separate forms varying in size, speed and power to act. It may be the next or even the final step in the development of the evolution of the cosmos.

We should not feel degraded by these, our offspring. Think how far we have come. Four billion years ago the Earth's surface was probably an ocean rich with organic chemicals. It was warm and comfortable and, at the time, did not need regulation from Gaia. Somehow, life started. The first forms of life were simple cells filled with chemicals. Gradually they took form and became what we might recognize as bacteria. These bacteria were alive and did not hesitate to hunt, kill and eat one another.

This went on slowly and stably for several billion years until about a billion years ago, when one of the bacteria that had been eaten survived within its predator and somehow the two living organisms formed a new life, the eukaryotic cell. From these the kingdoms of plants and animals evolved. The bacteria and other unicellular organisms remained and played their part in making a living planet. This outstanding biological discovery was made by Lynn Margulis who I am glad to say was a close friend and colleague for many years.

Then, with the appearance of humans, just 300,000 years ago, this planet, alone in the cosmos, attained the capacity to know itself. Not at once, of course; it was not until the appearance of the titans of the scientific renaissance a few hundred years ago that humans began to grasp the full physical reality of the cosmos. We are now preparing to hand the gift of knowing on to new forms of intelligent beings.

Do not be depressed by this. We have played our part. Take consolation from the poet Tennyson writing of Ulysses, the great warrior and explorer, in old age:

Tho' much is taken, much abides; and tho'
We are not now that strength which in old days
Moved earth and heaven; that which we are, we are . . .

That which we are, we are. That is the wisdom of great age, the acceptance of our impermanence while drawing consolation from the memories of what we did and what, with luck, we might yet do. Also, perhaps, we can hope that our contribution will not be entirely forgotten as wisdom and understanding spread outwards from the Earth to embrace the cosmos.

Acknowledgements

Like a newly designed and built plane, a book can fly serenely or fail to leave the ground and remain unread.

This book has benefited from the wisdom and craftsmanship of Bryan Appleyard and Stuart Proffitt. With such help it was naturally airworthy, and as I watched it rise gracefully into the stratosphere where good books fly I felt a sense of deep gratitude to them both. The conception of the book was shared by my dear wife, Sandy, and nurtured as it progressed by scientist colleagues, especially the Astronomer Royal, Martin Rees, and by my good friends Bruno Latour, Tim Lenton and the late Lynn Margulis.

Index